GENDER AND SECURITY IN DIGITAL SPACE

Navigating Access, Harassment, and Disinformation

*Edited by Gulizar Haciyakupoglu
and Yasmine Wong*

Routledge
Taylor & Francis Group

LONDON AND NEW YORK

Cover image: Getty Images

First published 2023
by Routledge
4 Park Square, Milton Park, Abingdon, Oxon OX14 4RN

and by Routledge
605 Third Avenue, New York, NY 10158

Routledge is an imprint of the Taylor & Francis Group, an Informa business

© 2023 selection and editorial matter, Gulizar Haciyakupoglu and
Yasmine Wong; individual chapters, the contributors

British Library Cataloguing-in-Publication Data
A catalogue record for this book is available from the British Library

ISBN: 978-1-032-19959-7 (hbk)
ISBN: 978-1-032-19958-0 (pbk)
ISBN: 978-1-003-26160-5 (ebk)

DOI: 10.4324/9781003261605

Typeset in Bembo
by KnowledgeWorks Global Ltd.

CONTENTS

LIST OF FIGURES

CONTRIBUTORS

Gabrielle Bardall specialises in feminist democratisation processes. She has served as advisor and educator to parliamentarians and electoral commissions in over 60 countries worldwide for nearly two decades. Dr. Bardall is Vice-President of External Relations for the Parliamentary Centre and a non-residential fellow with the Centre for Democracy and Technology. Her research on violence against women in politics has received numerous awards. She holds degrees from the Université de Montréal, Science-Po Paris and McGill University.

Gulizar Haciyakupoglu is a Research Fellow at the Centre of Excellence for National Security (CENS) of the S. Rajaratnam School of International Studies (RSIS), Nanyang Technological University, Singapore. Her research concerns online harms, trust, and activism. She holds a PhD from the National University of Singapore (NUS), Communications and New Media Department (CNM), and an MA in Political Communication from the University of Sheffield. She received her bachelor's degree in Global and International Affairs from the Dual-Diploma Programme of the State University of New York (SUNY) Binghamton, and Bogazici University, Turkey.

Sun Sun Lim is a Professor of Communication and Technology and Head of Humanities, Arts and Social Sciences at the Singapore University of Technology and Design. She has extensively researched the social impact of technology, focusing on technology domestication, digital disruptions, and smart city technologies. She recently published *Transcendent Parenting – Raising Children in the Digital Age* (Oxford University Press, 2020) and co-edited *The Oxford Handbook of Mobile Communication and Society* (Oxford University Press, 2020). She serves on 11 journal editorial boards. From 2018 to 2020, she was a Nominated Member of the 13th Parliament of Singapore. She was named to the inaugural Singapore

100 Women in Tech list in 2020 for her pioneering research on the social impact of technology. She frequently offers her expert commentary in diverse outlets including *Wall Street Journal, Guardian, La Tribune, Nature, Scientific American,* Channel NewsAsia, *Business Times, and Straits Times.*

Priyank Mathur is Founder and CEO of Mythos Labs, a social enterprise that uses media and technology to counter gender inequality, violent extremism, and misinformation. On behalf of clients including UN Women, the US Department of State, and the European Commission, Priyank has led projects in 14 countries as well as authored multiple reports researching online misogyny and hate speech. He is host of acclaimed documentaries and podcasts including the United Nations' ExtremeLives and Comedians Explain the World. His work has been profiled on BBC, CNN, the Singapore Ministry of Home Affairs' *Home Team Journal*, Bangkok Post, and India Ahead TV. Previously, Priyank served as a Policy Analyst and Intelligence Officer at the US Department of Homeland Security. He also served as Global Consulting Director at Ogilvy and as a Contributing Writer at The Onion. Priyank holds an MBA from MIT Sloan School of Management, as well as an MA and BA in International Relations from Boston University.

Tamara Nair is a Research Fellow at the Centre for Non-Traditional Security Studies (NTS Centre) at the S. Rajaratnam School of International Studies (RSIS), Nanyang Technological University. Tamara's research focuses on issues of Power and the biopolitics of female labour, and the Women, Peace and Security (WPS) agenda in Southeast Asia. Her current research interest specifically addresses women's security in digital space. She is Singapore's representative of the ASEAN Women for Peace Registry and has authored the 2018 Human Rights and Peace Education Report for Singapore. She is also the representative for Nanyang Technological University for the ASEAN University Network on Human Rights and Peace Education. She has published in Development Studies journals; writing on marginalised communities and sustainable development, issues of gender, power, and subject creation.

Sarah Shoker is a Research Scientist at OpenAI, where she focuses on the geopolitics of artificial intelligence. She was previously a SSHRC postdoctoral fellow in political science at the University of Waterloo, where she worked on the policy impact of emerging technologies on international security. She was also the winning beneficiary for the 2019–2020 University of Waterloo Trailblazer Postdoctoral Fellowship. She completed her PhD at McMaster University.

Fitriani Bintang Timur is a Researcher at the Department of International Relations, Centre for Strategic and International Studies (CSIS). Broadly, her research focus includes women in peace and security, peacekeeping, peace and mediation, as well as cyber security. Fitriani is also a lecturer at the Department of International Relations at Universitas Indonesia, teaching for modules on

Transnational Community, Transnational Network and Gender in Global Affairs. She is a co-editor of a book titled *"Towards a Resilient Regional Cyber Security: Perspectives and Challenges in Southeast Asia"* (2019). She obtained her PhD in Security and Defence Studies from Cranfield University, United Kingdom.

Kristina Wilfore is a seasoned international development and campaigns and elections professional who has worked in over 25 countries for more inclusive and responsive democratic movements. She has been on the ground in hotspots such as Ukraine, Kenya, Turkey, Afghanistan, Kosovo, and Syria. Kristina has worked hand-in-hand with hundreds of women on their campaigns for higher office and to help break systemic barriers to political participation. Along with Lucina Di Meco, she co-founded #ShePersisted, a cross-national initiative to tackle gendered disinformation and online attacks against women in politics. She serves as an Adjunct Professor with The George Washington University's Elliott School of International Affairs, teaching a course in countering disinformation and is a member of the US Institute of Peace Civil Society Working Group.

Yasmine Wong is a Senior Analyst with the Centre of Excellence for National Security (CENS) of the S. Rajaratnam School of International Studies (RSIS), Nanyang Technological University, Singapore. Her current research focuses on issues pertaining to social resilience, social cohesion, and inter-group relations in online and offline spaces. Yasmine received her bachelor's degree in Social Sciences (Politics and Philosophy) from the University of Manchester, and holds a Masters of Science in Political Sociology from the London School of Economics and Political Science.

FOREWORD

In May 2021, the High Commission of Canada in Singapore and the S. Rajaratnam School of International Studies jointly organised a webinar series under the theme "Gender, Security and Digital Space: Exploring Risks, Opportunities, and Security Implications". I was honoured to participate, underscoring Canada's commitment to addressing issues related to Women and Cyber and exchanging best practices with a Southeast Asian audience and beyond. This is a new and evolving threat environment, and Canada seeks to address the serious security threats of online harm against women and provide potential solutions, through both research and practical initiatives. While there is no doubt digital technology and the Internet have the potential to advance human rights, gender equality, and inclusion, they can also breed exclusion and facilitate gender-based violence, harassment, and threats.

During the COVID-19 pandemic, there has been an increase in violence against women globally, including technology-facilitated sexual violence, exacerbated by the proliferation of information manipulation online. The timing was opportune to elevate awareness of gender equality issues online by framing the discussion from a much-needed security lens.

The experts featured in the webinar series argued that the security implications of misogyny online are vast and serious, causing major harm not only to women, but also to our democracies. They highlighted the following key takeaways:

- Studies in multiple countries have pointed to **gendered disinformation** as a tried and tested tactic employed by state and non-state actors. Coordinated and targeted campaigns leverage misogyny, patriarchy, and hate, in order to silence opposition voices, erode trust in democratic processes, and undermine democratic principles.

- This challenge is **intersectional.** Women of colour often face a greater level of online abuse. By harassing, spreading hate, threatening rape and violence, often through racist and racialised, sexual and transphobic narratives, perpetrators seek to dehumanise, discredit, and intimidate women. There is little to no accountability for perpetrators and online platforms that carry such content.
- As a result, women and other groups facing gender discrimination often begin to **limit their online interactions or disengage completely,** and forfeit their right to freedom of expression by leaving platforms entirely. The chilling effect also results in **deterring women from participation in public life**, as journalists, politicians, human rights defenders, etc.

Global Affairs Canada's Centre for International Digital Policy has undertaken research on anti-Asian narratives in the COVID-19 context. This research pointed to the weaponisation of gender-based narratives, such as extremist voices leveraging gender to spread hate. What we need now, more than ever before, is a democratic vision for the Digital Age. How do we continue to promote democracy, human rights, and gender equality as life increasingly shifts online?

Canada's response is **digital inclusion**, which refers to the full and meaningful access to and use of digital technology and the internet for all. It comprises not only availability of physical infrastructure for connectivity and access to the intangible elements of connectivity, but also online spaces conducive to civic participation free from censorship, and the ability to trust industry and governments to keep users safe from online harms – such as gender discrimination – and violations of privacy.

Canada champions digital inclusion at home through countless initiatives, beginning with our Digital Charter. For example, the *Digital Citizen Initiative* is one of our flagship programmes to build public resilience to online disinformation. Canada also promotes digital inclusion globally:

- We produced a *Playbook for Gender Equality in the Digital Age*, which sets forth best practices to support gender equality in digital contexts.
- We support the Women in Cyber fellowship programme to support the participation of female diplomats from the Global South in UN cyber Open-Ended Working Group negotiations.
- We lead the G7 Rapid Response Mechanism aimed at identifying and responding to foreign threats to democracy, including foreign state-sponsored disinformation. In this context, we consider how gender-based disinformation is leveraged for the purposes of foreign interference.
- We are an active member of the Freedom Online Coalition, which we will chair in 2022, where we press for digital inclusion including to address online violence and discrimination against women and girls; during our 2022 Chairship, we will promote evidence-based policymaking to address online gender-based violence through research and advocacy efforts.

- Canada also co-champions the UN Secretary General's roundtable on digital inclusion and continues to empower women via ICT access and skills-building.

I hope that this book will be a valuable resource to help decision-makers, including governments and social media platforms, to understand the security imperative of addressing the treatment of women online. At a moment of rapid pandemic-accelerated digitalisation, governments worldwide, including in Southeast Asia, are struggling to safeguard the integrity of their information space, promote cybersecurity, and guard against foreign interference. The online space is not gender neutral. Absent diverse perspectives and approaches, the cyber domain will not be safe for anyone.

Gallit Dobner,
Director, Centre for International Digital Policy
of Global Affairs Canada

ACKNOWLEDGEMENTS

The idea of this edited volume came about during the planning phase of the three-part "Gender, Security and Digital Space" webinar series co-organised by the Centre of Excellence for National Security (CENS) at the S. Rajaratnam School of International Studies (RSIS) and the Canadian High Commission (CHC) in Singapore. We, the editors of this volume, and Benjamin Ang from CENS met Toby Schwartz, who was then the Counsellor of Political and Public Affairs at CHC Singapore, at a cafe in mid-March to sketch out the plan of the webinar series. As we negotiated priorities and tried to draw the scope of the webinar series, we identified a shared concern: we did not want the discussions on gender, security, and digital space to remain in talks. We wanted the webinar series to initiate conversations that would later be captured and expanded on in written work. Fast forward a year, and here we are presenting readers with this edited volume.

We would like to thank CHC (especially Toby) and CENS for their support.[1] Benjamin Ang from CENS deserves a special, heartfelt thank you. He has been a part of this project since conception and continued to support it in every step of the way. He spared time to be our sounding board, provided advice whenever we needed, and shared his opinion on parts of this work. We would like to thank him for setting a great example for us with his professionalism.

This volume would not have come to fruition without the dedication of contributing authors. We would like to thank them for the time they invested in multiple rounds of revisions, despite their busy schedules. We also would like to express our gratitude to the anonymous reviewers of our edited volume proposal. Their feedback provided invaluable help in fine-tuning the framework of this project. And of course, a special thanks to the Routledge team for helping us throughout the project.

We would also like to thank our families for their support and love. Lastly, this journey has been as painless as editing journeys go – and we have each other to thank for that.

Note

1 We did not receive any financial support from the High Commission of Canada for this edited volume.

LIST OF ABBREVIATIONS

ACWC	ASEAN Commission for the Promotion and Protection of the Rights of Women and Children
AMS	ASEAN Member States
ASEAN	Association of Southeast Asian Nations
AWARE	Association of Women for Action and Research
BART	San Francisco Bay Area Rapid Transit System
BN	Barisan Nasional
CEDAW	Convention on the Elimination of All Forms of Discrimination against Women
CEO	Chief Executive Officer
CIO	Chief Information Officer
CSO	Civil Society Organisations
DoT	Daughters of Tomorrow
DVA	Domestic Violence Act
EU	European Union
GBA+	Gender-Based Analysis Plus
GBV	Gender-Based Violence
GBVO	Gender-Based Violence Online
GDPR	General Data Protection Regulation
HR	Human Resources
HTML	HyperText Markup Language
ICT	Information and Communication Technologies
IFES	The International Foundation for Electoral Systems
ISIS	Islamic State of the Iraq and the Levant
ISO	International Organization for Standardization
IT	Information Technology
ITU	International Telecommunication Union

IWCS	Indonesia's Women in Cybersecurity
LGBTQ+	Lesbian, Gay, Bisexual, Transgender, and Queer/Questioning
LGBTIQ+	Lesbian, Gay, Bisexual, Transgender, Intersex, and Queer/Questioning
MAU	Monthly Active Users
MP	Member of Parliament
MSC	Multimedia Super Corridor
NAPs	National Action Plans
NGO	Non-Governmental Organisation
NUS	National University of Singapore
OEWG	Open-Ended Working Group on Digital ICTs in the Context of International Security
RPAEVAC	Regional Plan of Action on the Elimination of Violence Against Children
SACC	Sexual Assault Care Centre
SDG	Sustainable Development Goals
SG	Singapore
SIM	Subscriber Identity Module
SIS	Sisters in Islam
SOGIE	Sexual orientation and gender identity and expression
SPF	Singapore Police Force
STEM	Science, Technology, Engineering, and Math
UK	United Kingdom
UN	United Nations
UNDIR	United Nations Institute for Disarmament Research
UNECE	United Nations Economic Commission for Europe
UNFPA	United Nations Population Fund
US	United States
USD	US Dollar
VAW	Violence Against Women
VAWG	Violence Against Women and Girls
VAWIE	Violence Against Women in Elections
VAWP	Violence Against Women in Politics
VPN	Virtual Private Network
WISE	Women in Standardization Expert Group
WPS	Women, Peace and Security

1

INTRODUCTION

Gulizar Haciyakupoglu and Yasmine Wong

A woman argued that her avatar was virtually gang raped in Meta's Horizon Venues.[1] As per her account, when she tried to escape, the perpetrators called out – "don't pretend you didn't love it" and "go rub yourself off to the photo".[2] Another woman shared how an avatar in the virtual shooting game Population One "simulated groping and ejaculating onto her avatar".[3]

The growing hype around the metaverse drew attention to such accounts of harassment in virtual reality. However, neither harassment and safety concerns, nor the shortfalls in action against bad behaviour in virtual reality spaces are new.[4] The recent examples from the metaverse pile onto the persistent problem of online harms in digital spaces, and just as in any digital space, they underscore fissures along gendered lines that fester gender-based toxicity. The developments in the metaverse along with the advancements in related technology, such as the elevation of the users' senses during virtual reality experience via gadgets like haptic gloves,[5] will likely expand the threat landscape and have gendered implications.[6] Besides, as observed in other technologies, the lack of access to the metaverse will spawn its own set of concerns.

The threats in digital space with gendered implications will continue to cause distress, as many countermeasures remain insufficient or lack a gender angle, and legislation and infrastructure play catch up with technological developments. Yet, growing insecurities in digital spaces and beyond should not be a cause to wallow in dystopianism, as technological advancements also entail novel opportunities and benefits to reap. The question is how the equalising and liberatory potential of digital spaces and technologies can be extracted without essentialising technology as unproblematic or as the only nostrum, and while curbing the harms that come with these developments. This edited volume engages with this inquiry with a focus on the persistent problems of access, harassment,

DOI: 10.4324/9781003261605-1

and disinformation through a gender lens, and with that, contributes to the underexplored intersection of gender, security, and digital space.

The combinations of gender, security, and digital space (and Information Communication Technologies (ICTs)) have long been a rich exploration ground for scholars and practitioners. The expansion in literature exploring the interaction between these concepts – notably, the nexus of gender and security, and of gender and digital space – is a nod to the desire to untangle unresolved debates on power relations, inequalities, and risks. These explorations stand against the rising tide of fresh issues, creating new research gaps to fill. Research that marries all three concepts – gender, security, and digital space, on the other hand – remains inadequately explored because of the dynamism of the security and technological landscapes, and inadequate attention to gender in relevant inquiries.

This edited volume explores this very triad through chapters by scholars and practitioners from various disciplines, and with that, provides for a mix of academic analyses and practitioner experiences. Within this intersection of gender, security, and digital space, sections of this volume zoom into issues surrounding access, harassment, and disinformation. On a thematic level, *justice and responsibility, equality in representation, and freedom of speech while being free from harms* are three tropes that run across chapters. Following this opener, we visit preceding literature on the topic. Subsequently, we delve into the content of this volume, with references to overarching themes, and short summaries of chapters.

From combinations to intersection: Gender, security, and digital space

The literature exploring the interaction between gender and security, and gender and digital space remains formative to the gender, security, and digital space agenda. Hence, we will provide a brief overview of the main discussions within, firstly, the intersection of gender and security, and secondly, within gender and technology – and digital spaces – before our foray into combining the three concepts.

Research at the intersection of *gender and security* has attempted to introduce gender in fields dominated by masculine norms. This is in line with the practices of problematising traditional understandings of security as a masculine concept and examinations of the marginalisation of women and their experiences in processes of security. Some feminist perspectives have drawn attention to the centrality of human subjects in International Relations,[7] making visible the lives of women in patriarchal systems of politics and militarism,[8] and illuminating the primacy of their needs in pursuits of security, thus challenging the state-centrism that dominates the field.[9] Other studies have examined how gendered assumptions motivate popular representations of men and women in political violence and war.[10] What follows is the essentialisation of women as victims, the lack of recognition of male victims, and the downplaying of female agency in conflict and post-conflict settings.[11]

In practice, the United Nations (UN) has advocated for gender–sensitive approaches to centre women's needs and perspectives in decisions surrounding peace and security, particularly with the Security Council Resolution 1325, which recognises the disproportionate effects of conflict on women and girls, and their unique contribution to conflict resolution and prevention.[12] The legacy of this endeavour is reflected in the subsequent resolutions on Women, Peace, and Security (WPS), which stress the significance of having women in decision-making positions, tackling sexual violence, understanding the gendered dimensions of conflict, and so on.[13]

Those venturing into the intersection of *gender and technology*, including cyberspace, on the other hand, have explored technology with a gender focus, especially women's interaction with technology and cyberspace. Mirroring fragmentations within the concept of feminism,[14] some scholars emphasised *cybernetic's* liberatory potential and the challenge it can potentially pose to patriarchal dominance,[15] while others warned that technology and cyberspace would not grant an easy "liberation".[16] The interrogation of concepts of power, identity, and autonomy, and the impact of new technologies on these concepts have been central to these debates.[17] Donna Haraway in her renowned "A Cyborg Manifesto" imagined a more pluralist future through cybernetics, where the obscured boundary between organism and machine (here she uses the term "cyborg") could serve as a vehicle for women to transcend traditional gender roles built on naturalised understandings of gender and bodies.[18] Sadie Plant observed that the "network culture still appears to be dominated by both men and masculine intentions and designs. But there is more to cyberspace than meets the male gaze".[19] Others were more sceptical. The claims that ground the persistence of male dominance in electronic networks included women constituting the minority of users and the dominance of masculine norms online.[20] Along these lines, Margie Wiley suggested that the Internet, with its lineage tracing back to "the military-industrial complex" and "academic institutions", has a "problematic relationship with gender".[21] Judith Squires, on the other hand, drew attention to the "exploitative and alienating potential of technology"[22] and warned against taking "an apolitical stance with regard to the form and operation of developing technologies".[23] To Squires, cyberfeminism requires a view of "cyborgs that may help to heighten our awareness of the potentiality for new technologies, and the political pursuit of the practical manifestation of an appropriate technology: unintimidating, accessible, and democratic".[24]

The "appropriate technology", however, remains an ideal, as the fundamental question of access continues to occupy agendas with the perpetual under-representation of women in digital spaces and industries. The digital divide curtails women's engagement in political and civil life,[25] and impedes women from accessing new markets, education, health and financial services, career opportunities, and decision-making in industry and politics.[26] Where access is a major concern, online security concerns and their gendered implications may face the risk of being side-lined in political deliberations. There is inertia in moving the conversation beyond strengthening women's participation in science, technology, engineering, and mathematics (STEM), and cybersecurity[27] into considering gendered implications of inequalities and threats emerging from newer ICTs. As Sarah Sobieraj (2020) aptly argues, "newer ICTs, like their predecessors, are new arenas in which gender inequalities are recreated, reformed, and resisted".[28] This suggests the need to couple access with gendered considerations of safety and security in cyberspace, and responsible uses of ICTs.

To address this growing landscape of online harms and security concerns, some studies invest attention to the areas at the intersection of *technology (and cyberspace) and security*, such as burgeoning research on disinformation campaigns,[29] albeit often without a specific focus on gendered implications. Yet, the recent increase in articles[30] and webinars[31] on issues surrounding gender and online harassment and disinformation signals the appetite for gender-focused inquiries into the intersection of *gender, security, and technology*, including those concerning digital spaces. Within this line of research, some of the works on online spaces contribute to the existing body of research on violence against women[32] and hint at the perseverance of gender-based harassment, disinformation, and other abuses across spaces and time.

Online gender-based harassment comes in many forms, including sexual harassment, privacy invasion, hate speech, and death or rape threats.[33] Gender-based online attacks build upon gender stereotypes, misogyny, and inequalities ingrained in societies.[34] As such, Sobieraj argues that approaching digital attacks through an analysis of power and inequality illuminates why identity-based attacks disproportionately affect women.[35] This parallels other feminist scholarship in their understanding of digital misogyny as another facet of the chronic bid to prevent women and other marginalised groups from equal participation in public deliberation.[36] Furthermore, the effects of digital attacks on women are jarring – encompassing a wide range of consequences, including economic, professional, psychological, and social damages.[37] Identity-based online attacks also have broader implications on political participation and democracy.[38] Among others, they pave the way for "patterned deterioration of political discourse", take a toll on the "robustness of [...] pool of candidates willing to run for public office"[39] and where election-related disinformation is concerned, challenge election integrity, and "undermine[] the role of journalism in a healthy democracy".[40] This is especially concerning as various studies warn that female politicians, journalists, and public figures are subject to online violence and gendered disinformation,[41] of which Nobel Peace Prize laureate Maria Ressa and US Vice President Kamala Harris are notable examples.

Harassment and disinformation in, as well as access to, online spaces, as demonstrated by the research above, remain unresolved issues jeopardising safe and civil engagement in cyberspace across genders. Inquiries into each of these aspects sit within the intersection of gender, security, and digital space, albeit often in isolation. Besides, the security dimension remains an under-emphasis in these ventures. This edited volume seeks to address this dearth in knowledge through its exploration of issues concerning gender, security, and digital space. Within this triad, chapters in this volume zoom into issues of access, harassment, and disinformation. Access, harassment, and disinformation are explored in separate sections, yet each chapter references more than one of these issues.

The relation between access, harassment, and disinformation is a complex one. Lack of, or insufficient, access to digital technologies impede victims' access to critical information and support when faced with violence.[42] It also exacerbates inequalities which surface from the unequal access to economic, social, political, and other opportunities provided by the Internet as well as ICTs.[43] Women need to have equal access to digital spaces and decision-making processes to arrive at inclusive solutions to online harms and inequality in access. On the other hand, as others also argue, access to digital space obliquely brings about the risk of exposure to online harassment and disinformation, which may curtail women and marginalised groups' reach in public and political spheres. Moreover, the price and quality of access,[44] the type and condition of the available hardware[45] for connection, access to and knowledge on available software for protection, and other factors create different online experiences. For online experiences to be safe and civil endeavours, equal, dependable, and safe access has to be coupled with digital literacy, and security.

Themes running throughout the edited volume

Our analysis of the chapters reveals three recurring themes that run across the chapters: justice and responsibility, equality in representation, and freedom of speech while being free from harms.

Justice and responsibility

There is a lack of a shared understanding[46] of what safety from online harms means and what equal access encompasses for different stakeholders. This leads to fractured notions of what is just when it comes to whose responsibility it is to secure women from online harms and of what is just when conceptualising security itself. How can a fair share of responsibility be achieved when this lack persists? Faced with this conundrum, the search into who is responsible for online security, as well as the need to consider diverse understandings of security from online harms, surfaces underlying assumptions and power structures that hinder the pursuit of gender equality. The issues surrounding responsibility, which also occupy the agendas of other scholars working on intersecting concerns,[47] transpire as a burgeoning, unresolved question in this volume.

Discussions on responsibility within the edited volume stand amidst a backdrop of existing debates that seek to locate where responsibility for online (in) security lies.[48] Here, accounts on responsibility and countermeasures concerning gendered threats reveal a plethora of views that seek to problematise the relation between (or find balance amongst) measures that foreground structural change (including those undertaken by governments), the accountability of social media platforms, civil society efforts, and individual responsibility. The perspectives on the responsibility of state-based actors vary across chapters. In most of the chapters, the state was brought into the picture as one of the actors responsible to create conditions conducive for women's security in digital spaces. Yet, states may as well undermine or fail to address the security of their own people and overlook gendered needs in some cases (Chapters 2, 3, 5, 6), and gendered disinformation and other online harms may be employed by illiberal state actors (domestic or foreign) against women in politics (Chapters 3, 7, 8). On the other hand, some chapters also shine a light on civil society and societal efforts – from instilling women with the skills to navigate digital spaces, to advocating for the adoption of a gendered lens in national security agendas (Chapters 3, 5, 6, 8–10). However, individual and civil society practices of security should not foreclose state and platform responsibility. On the latter, some chapters discuss the importance of holding tech and social media companies accountable (Chapters 8–10), amidst scepticism over the willingness of these companies to take action.

What results is the examination of individual responsibilities in tandem with the need for accountability from parties in power. More specifically, this demands whole of society approaches, and the cultivation of the will to avoid further burdening of victims and those vulnerable to the threat of online harms. Some see information literacy as a fundamental component of the solution and present successful examples (Chapters 9 and 10). Nevertheless, information literacy initiatives do not offer a panacea, as, as Wilfore argues in Chapter 8, they may have limited effect over the emotional sway that misogynistic content has over individuals with implicit biases against women.

Lastly, some chapters turn to intergovernmental organisations, and/or the cultivation of transnational cooperation when exploring issues and their solutions (Chapters 2–4, 8). More specifically, some authors refer to their definitions (Chapters 2, 4, 5, 7, 8) or frameworks and/or resolutions (Chapters 2–4, 8) when building context or proposing policy considerations. A question to explore is whether intergovernmental organisations can take the lead in forging shared language and frameworks that are context, culture, and gender–sensitive.

Freedom online; freedom from harms

Chapters in this volume seek to balance between the mitigation of risks to women in digital spaces and the "liberating" potential of the Internet and ICTs. This is particularly pronounced with regard to freedom of speech and security from harms. Along these lines, various authors point out the freedoms and benefits afforded by cyberspace and ICTs, such as the economic and social opportunities digitalisation offers (Chapters 3, 4, 9), and space for advocacy and non-elite practices of security on social media (Chapters 5 and 6). Yet, many authors also speak on the threats to freedoms, including freedom of speech, in online spaces. They call attention to how digitalisation facilitate online gender-based violence (GBV) (Chapters 2, 4–10), and how toxic discourses and behaviours, as well as control over political speech and Internet access, may threaten the security of or silence women online (Chapters 3, 5, 6). These views overlap with some arguments in literature, which point out that online toxicity may lead women and marginalised groups to self-censor and exclude themselves from participating due to fears of abuse.[49]

The review of the chapters in this volume also suggests that security and freedoms, including freedom of speech, cannot be seen as negatively correlated – this rests on limited understandings of both. Instead, security and online freedom, including freedom of speech, should have inclusive and considerate definitions. With specific regard to gender, this calls for an understanding of online freedoms in a way that does not infringe the rights of women and marginalised groups' access to and safety in online spaces.

Equality in representation

Lastly, at the heart of the issue, there is consensus among most authors on the importance of equality of representation and the pervasive impact that the lack of non-tokenistic representation has on all levels of technological design, management, and solutioning processes. Genuine representation of women would, among others, reduce the likelihood of essentialising gendered categories, for example, of women as necessarily vulnerable (Chapter 2), or of victims as women (Chapter 6). Some chapters discuss the issue of representation at the state level, where the lack of gendered representation affects the quality of democratic institutions, and where ensuring that channels for women to participate in public deliberation and public life are safe is paramount (Chapters 3, 5, 6, 8). More specifically, attempts at defining online security threats should involve consultation of women and their experiences of technology and cyberspace to avoid overlooking the insecurities they face (Chapters 5 and 6). Others call for their greater participation in STEM, the technology industry, and more specifically in cybersecurity (Chapters 2–4, 9). Although equal gender representation has not been actualised in diplomatic, humanitarian, and security arenas, there is increasing recognition of women's contributions to effective conflict resolution and peace,[50] and some (Chapters 2 and 3) offer this as a promising example that can be replicated across different disciplines. Besides, the incorporation of women's voices at all levels would help prevent top-down decision-making that may overlook or misrepresent the needs of women. Attempts by some authors to tackle unresolved tensions emerging from traditional power structures, many of which privilege patriarchal worldviews, could provide guidance in this endeavour.

A taste of what's to come

We deliver the chapters[51] under four sections: (a) Old threats, new spaces; (b) Gender, online harassment, and the security question in Southeast Asia; (c) Gendered disinformation and violence in digital space(s); and (d) Countermeasures in place.

a. The first section, "Old threats, new spaces", unites two chapters that marry existing and looming security matters and concerns related to gender inequality.

   Tamara Nair opens the section with a chapter on gendered inequalities in digital space with a focus on issues and concerns that have been "invisibilised". Through a framework of ignorance as politically produced non-knowledge, Nair explores how various exercises of power and notions of vulnerability curtail some groups and individuals' access to political agency in cyberspace. Adapting existing frameworks to new spaces, Nair suggests the adaptation of UN Security Council Resolution 1325 – the WPS agenda to address gender inequality and violence against women – in digital spaces.

   Following Nair's exploration of the ignorance of gendered vulnerabilities in digital spaces, Sarah Shoker drills down to the specifics, outlining the disproportionate impacts of Internet shutdowns on women. This remains a lesser-explored area, perhaps because Internet shutdowns can be initiated by states - often in the name of securing turbulent political environments –rather than a form of cybercrime, although Internet shutdowns create conditions for the backsliding of women's equality. Shoker advocates for the collection and analysis of gender-specific statistics to better capture women's experiences with digital ICTs as well as to understand and explicate the positive relationship between women's rights and national security.

   These two chapters approach the question of access from two different angles. However, both authors assert that the issues they unpack have gendered impacts, yet the gender aspect is missing from the policies and solutions that target the problems they discuss. Both cite intergovernmental organisations when discussing the policies and solutioning processes. Both refer to the relationship between gender inequality, gendered abuse and harms, and international peace; and both highlight the need to aggregate gender-focused data. Their discussions, especially their references to intergovernmental organisations and their frameworks, invite further research on the interaction between intergovernmental and national policies and how it relates to individual insecurities. They also raise questions on the division of responsibility as well as the question of enforcement on issues debated at the intergovernmental level.

b. The second section, "Gender, online harassment, and the security question in Southeast Asia", focuses on issues concerning harassment in digital spaces. It explores the security conundrum surrounding online harassment and Internet's impact on gender equality with attention to the Southeast Asian

(SEA) context. While there is burgeoning literature on online harassment, focus on the SEA remains limited and this section addresses this gap.

Fitriani Bintang Timur starts this section with a chapter that focuses on (digital) conditions in the SEA region, where she believes digital transformation is upheld as a key to economic development while awareness on gendered harms in cyberspace remains inadequate. Departing from United Nations Institute for Disarmament Research's (UNIDIR) Three Pillar Framework, Fitriani explores Association of Southeast Asia Nations (ASEAN) member states' response to GBV and postulates the need for countermeasures against the threat and for a gender–sensitive approach.

In the next chapter, Gulizar Haciyakupoglu probes how the limitations to the space for political speech might create insecurities for gender equality advocates in Malaysia, especially when online harms challenge civil and secure conversation online. She invests particular attention to how the justification of protecting perceived social cohesion and "Asian values" might curtail the space for political speech, and questions if this approach could create insecurities for gender equality advocates and women in general. How the limitations as well as protections afforded by the legal framework influence gender equality advocacy in Malaysia is also central to her inquiry. Haciyakupoglu highlights the need to negotiate what gendered online harassment constitutes for different stakeholders in Malaysia and the need to expand the space for political speech.

In the final chapter of the section, Yasmine Wong discusses the phenomenon of informal pursuits of justice online by survivors of sexual assault and harassment through the framework of vernacular security, highlighting the politics of emotions that undergird informal practices of security. The chapter pays particular attention to case studies from Singapore and discusses how women, through their expression of emotional personal narratives on social media, have defined and sought security amidst the threat of sexual harassment and assault, and the shortfalls of current means of recourse, which underscores the value of emotions in digital spaces. The resultant effect – the problemising of traditional, state-centric, and gendered notions of security, agency, and power.

The chapters in this section not only provide a SEA perspective in Western-dominated fields, but also put forth different interpretations of the relationship between gender, security, and digital spaces. While violence against women (both in online and offline spaces) is central to these chapters, Fitriani situates her analysis within state-based (and inter-state) solutions to countering the problem, while Haciyakupoglu and Wong explore the idea that there is a fundamental need for state and civil society, or state and individuals to negotiate conceptions of (in)security. Haciyakupoglu and Wong also discuss the significance of digital spaces as arenas for women's participation in public deliberation, advocacy, and practices of security, and the importance of safeguarding these spaces against gendered abuse.

c. The third section, "Gender and Disinformation", delves into gendered implications of disinformation and abuse in digital space. Gendered implications of disinformation remain under-researched and ambiguities around terms and conflation of meanings impede effective communication and resolution of the problem.

Gabrielle Bardall, in the opening chapter of this section, situates "gendered disinforming" within violence against women in politics and clarifies misconceptions about disinformation that clouds the concept. The chapter invites more precise uses of the term.

Next, Kristina Wilfore contextualises gendered disinformation within GBV and argues that the term builds on "sexism, misogyny and the 'manosphere', 'online violence' and 'disinformation'". Wilfore suggests that understanding disinformation through a gender lens enhances our grasp of how sexist narratives are weaponised to intimidate women and impede their participation in public deliberation. Wilfore argues that enhancing political discourse to acknowledge the enabling role of technology, and social media as a behavioural modification system in attacks against women is essential for securing and furthering women's rights. It is also a crucial foreign policy and national security imperative for "democratic-minded" countries.

While authors in this part agree on some of the impacts of gendered disinformation, they show differences in their approaches to gendered disinformation, providing different perspectives on the issue. These two chapters invite an exploration of a problem, disinformation, that has often been explored without a gender perspective with a specific focus on its gendered implications.

d. And the fourth section, "Countermeasures in Effect", discusses remedies to access, harassment, and disinformation-related concerns. The two chapters in this section investigate efforts to combat online gendered disinformation, violence, and discrimination that are already in practice in different parts of the world.

First, Sun Sun Lim discusses the imperative to vest women with the knowledge of how to navigate digital spaces, and how they can help themselves and other women suffering from online harms. She examines notable digital literacy programmes in Singapore led by civil society groups and distils best practices with a view towards emulation by similar programmes in other countries.

Next, Priyank Mathur shares his experiences in the field as the Founder and CEO of Mythos Labs, a social enterprise that uses media and technology to counter harmful (and indeed gendered) narratives and disinformation. In his chapter, Mathur engages with original research that demonstrates the growing threat of online misogyny and hate speech against women in Asia and discusses the strengths and limitations of innovative countermeasures carried out by Mythos Labs across South Asia. The countermeasures include "My Power" trainings which focus on information literacy to combat online misogyny, as well as a campaign that uses humour to counter misogynist narratives.

Both authors discuss countermeasures targeted to equip individuals against online harms – this includes digital literacy and digital resilience programmes. Lim postulates that digital literacy programmes contribute to a longer-term goal of greater representation and gender equality in STEM industries and in decision-making positions as a sustainable way to introduce diversity into the design of technologies and security in digital spaces. Mathur warns that new modes of digital communication will give rise to new forms of gendered abuse and insecurities and argues for the need to adopt an approach that involves multi-stakeholder perspectives. Yet, new technologies and new modes of communication, as Mathur argues, will also provide new opportunities to design imaginative and successful programmes.

As portrayed by the chapter summaries, authors of this volume explore access, online harms, and security from different dimensions. Despite differences in focus and variance in views, there are overlaps in policy recommendations, including the need to: (a) expand research and gender-focused data-gathering efforts; (b) work on definitions and standards; (c) have women in decision-making and leadership roles; (d) understand the role of information (and digital) literacy; (e) raise awareness on the gendered implications of online threats; and (f) forge domestic and transnational collaborations.

Before we move on to the first section of the volume, we would like to offer some clarifications on the overarching terms gender, security, and digital space, and share our acknowledgements. On the concept of gender, the inclusion of women's experiences and perspectives in a field traditionally dominated by masculine viewpoints places the edited volume within the feminist agenda of embracing a gender–sensitive lens. Besides, some chapters within this edited volume touch on the need to problematise and re-examine masculine and patriarchal structures. More pertinently, the adoption of a gender-focused outlook in the field of security and technology invites attention to diverse experiences of (in)securities, threats, and solutions across the gender spectrum. Nonetheless, the chapters focus disproportionately on women, with some citing the primary concentration of available resources (literature and data) on women as a reason. Again, the scarcity of data and literature on multiple intersecting identities between demographic factors like gender, sexual identity, race, ethnicity, and socioeconomic status hinders the further examination of intersecting identities and the compounding of insecurities for some women and persons with different gender and sexual identities. Nonetheless, this volume acknowledges the importance of exploring gender beyond the binary (i.e., women or men), and with attention to diverse gender and sexual identities, and invites future research to attend to this gap.

With regard to security, the desire to explore what security means for different parties is palpable in each chapter. The chapters assume diverse approaches to the concept of security as each explores the concept in relation to a different topic – whether it is national or individual security. Each chapter expounds the authors' hopes for a more secure experience in and with digital space. As to digital space – and ICTs, authors in this volume base their inquiries on various online spaces and cite myriad technologies. However, although digital space takes centre stage throughout the volume, given the interconnectedness of virtual and physical worlds, online and offline lives are enmeshed in ways that are hard to entangle and problems in one extend into another. Not to mention, a long-term solution to ICT-facilitated VAW and *other issues* discussed in this volume requires the addressing of underlying problems, including gender inequality and misogyny. As such, chapters expose these intersections and continuities between the virtual and the physical, despite the volume's focus on digital space.

This edited volume, by virtue of its concerns with security, may give the impression of leaning into the negatives – that technology is tainted by pessimism if you will. This is a complication present not only in security-related literature, but also in works surrounding gendered violence in online spaces. Recognising this, we urge readers to keep in mind that the very tools that facilitate gendered harms, may also be used to counter them. As Sobieraj suggests, gender inequalities have been "recreated, reformed, and resisted" by new and old ICTs.[52] And after all, as James Bridle aptly states, when creating new tools and technologies, we project upon it a certain understanding of the world, the offloading of a particular thought or a way of thinking.[53] As such, instead of essentialising technology within binary views of – good and bad; benevolent and malicious – efforts should be invested in "re-enchant[ing]"[54] tools towards more equitable ends.

"Unintimidating, accessible, and democratic" technology continues to be an ideal, against the surge of online harms. This edited volume is a small step towards the ideal of a safe and civil cyberspace, and an invitation to build more knowledge for the road ahead. We envisioned this volume to not only contribute to a novel area that demands attention and research, but also serve as a digestible guidebook to practitioners and policymakers working on interesting topics. We hope we accomplished our aim.

Notes

1 Nina Jane Patel, "Reality or Fiction?", *Medium*, 21 December 2021, https://medium.com/kabuni/fiction-vs-non-fiction-98aa0098f3b0
2 Ibid.
3 Sheera Frenkel and Kellen Browning, "The Metaverse's Dark Side: Here Come Harassment and Assaults", *The New York Times*, 30 December 2021, https://www.nytimes.com/2021/12/30/technology/metaverse-harassment-assaults.html
4 Hannah Murphy, "How Will Facebook Keep Its Metaverse Safe for Users?", *Financial Times*, 12 November 2021, https://www.ft.com/content/d72145b7-5e44-446a-819c-51d67c5471cf; Tanya Basu, "The Metaverse Has a Groping Problem Already",

MIT Technology Review, 16 December 2021, https://www.technologyreview.com/2021/12/16/1042516/the-metaverse-has-a-groping-problem/; Jordan Belamire, "My First Virtual Reality Groping", *Medium*, 21 October 2016, https://medium.com/athena-talks/my-first-virtual-reality-sexual-assault-2330410b62ee

5 Lauren Goode, "Facebook Reaches for More Realistic VR with Haptic Gloves", *Wired*, 16 November 2021, https://www.wired.com/story/facebook-haptic-gloves-vr/

6 Yasmine Wong and Gulizar Haciyakupoglu, "Southeast Asia Must Be Wary of Gendered Cyber Abuse", *The Diplomat*, 13 June 2022, https://thediplomat.com/2022/06/southeast-asia-must-be-wary-of-gendered-cyber-abuse/

7 Aili Mari Tripp, Myra Marx Ferree, and Christina Ewig, eds. *Gender, Violence, and Human Security: Critical Feminist Perspectives*. New York: NYU Press (2013); Natalie Florea Hudson, *Gender, Human Security and the United Nations*, London: Routledge (2009).

8 Cynthia Enloe, *Bananas, Beaches and Bases: Making Feminist Sense of International Politics*. 2nd edition. California: University of California Press (2014).

9 Laura J. Shepherd, "Gender, Violence and Global Politics: Contemporary Debates in Feminist Security Studies", *Political Studies Review*, 2009, Vol. 7, p. 209; Cynthia Enloe, *Globalization and Militarism*, Plymouth: Rowman & Littlefield (2007).

10 Linda Åhäll, *Sexing War/Policing Gender: Motherhood, Myth and Women's Political Violence*, London: Routledge (2015).

11 Megan Mackenzie, "Securitization and Descuritization: Female Soldiers and the Reconstruction of Women in Post-Conflict Sierra Leone", *Security Studies*, 2009, Vol. 18, Issue 2, p. 18.

12 "Promoting Women, Peace and Security", *United Nations Peacekeeping*, https://peacekeeping.un.org/en/promoting-women-peace-and-security

13 Ibid.

14 Barbara M. Kennedy, "Introduction", in David Bell and Barbara M. Kennedy (eds.) *The Cybercultures Reader*, p. 283, New York: Routledge (2000).

15 E.g., Haraway and Plant.

16 Judith Squires, "Fabulous Feminist Futures and The Lure o Cyberculture," in David Bell and Barbara M. Kennedy (eds.) *The Cybercultures Reader*, pp. 360–373, New York: Routledge (2000); Kennedy, p. 285

17 Kennedy, p. 285.

18 Donna Haraway, "A Cyborg Manifesto: Science, Technology and Socialist-Feminism in the Late Twentieth Century," in David Bell and Barbara M. Kennedy (eds.) *The Cybercultures Reader*, pp. 291, 292–316, New York: Routledge (2000); Haraway in Kennedy, p. 285; Nicola Henry and Anastasia Powell, "Embodied Harms: Gender, Shame, and Technology-Facilitated Sexual Violence", *Violence against Women*, 2015, Vol. 21, Issue 6, p. 762.

19 Sadie Plant, "On the Matrix: Cyberfeminist simulations", in David Bell and Barbara M. Kennedy (eds.) *The Cybercultures Reader*, p. 325, New York: Routledge (2000).

20 Spender (1995) as cited in Nina Wakeford, "Networking Women and Grrrls with Information/Communication Technology: Surfing Tales of the World Wide Web", in David Bell and Barbara M. Kennedy (eds.) *The Cybercultures Reader*, pp. 350, 351, New York: Routledge (2000).

21 Margie Wiley (1995) as cited in Wakeford, p. 350.

22 Squires, p. 369.

23 Ibid, pp. 361, 369.

24 Ibid, p. 371.

25 Bardall in Segrave and Vitis, p. 119.

26 "Bridging the Gender Divide", *ITU*, November 2019, https://www.itu.int/en/media-centre/ backgrounders/Pages/bridging-the-gender-divide.aspx; Fitriani B. Timur, "ASEAN Gender Digital Disparities", Panel 1, 11 May 2021, in "CENS & The High Commission of Canada Webinar Series"; Haciyakupoglu and Wong, pp. 5–6.

27 Haciyakupoglu and Wong, pp. 5–6.
28 Sarah Sobieraj, *Credible Threat: Attacks against Women Online and the Future of Democracy*, p. 10, London: Oxford University Press (2020).
29 For example, there are organisations such as Graphika, Stanford Internet Observatory (Cyber Policy Center), and Digital Forensic Research Lab that publish reports concerning disinformation campaign on regular basis. Also examples of books discussing issues surrounding disinformation campaigns include (but not limited to) Sinan Aral, *The Hype Machine: How Social Media Disrupts Our elections, Our Economy, and Our Health – and How We Must Adapt*, New York: Currency (2020) and Thomas Rid, *Active Measures: The Secret History of Disinformation and Political Warfare*, London: Profile Books (2020).
30 Lucina Di Meco, "Gendered Disinformation and Online Attacks against Women in Politics", in *Gender, Security and Digital Space Webinar Series Panel 2*, 18 May 2021; Gabrielle Bardall, "Gender, Disinformation & Politics: Part of the spectrum of VAWP", in *Gender, Security and Digital Space Webinar Series Panel 2*, 18 May 2021; Nina Jankowicz, 18 May 2021.
31 Soraya Chemaly, Lucina Di Meco, Kristina Wilfore, Dhanaraj Thakur, and Allie Brandenburger, "Gender, Politics & Disinformation on Social Media" [Webinar], *Centre for Democracy & Technology*, 16 September 2020; Lucina Di Meco, Irena Hadžiabdić, Liubov Tsybulska, and Nina Jankowicz, "Gender Dimensions of Disinformation in Elections, Politics and the Digital Information Space" [Webinar], *International Foundation for Electoral Systems*, 24 February 2021; Maria Ressa, Alicia Wanless, and Shireen Mitchell, "At the Vanguard: How Women Lead the Charge in Researching Influence Operations" [Webinar], *Carnegie Endowment for International Peace*, 23 March 2021; Lucina Di Meco and Kristina Wilfore, "#ShePersisted: Gendered Disinformation Campaigns + the 2020 Elections" [Webinar], *Rep19*, 30 June 2020.
32 The works that look into gender based abuse, disinformation and/or hate speech in online spaces include: Sobieraj, 2020; Nina Jankowicz, Jillian Hunchak, Alexandra Pavliuc, Celia Davies, Shannon Pierson, and Zoe Kaufmann, Malign Creativity: How Gender, Sex, and Lies Are Weaponized against Women Online", The Wilson Centre, Science and Technology Innovation Program, January 2021; Lucina Di Meco, "Gendered Disinformation, Fake News, and Women in Politics", Council on Foreign Relations, 6 December 2019, https://www.cfr.org/blog/gendered-disinformation-fake-news-and-women-politics; Caroline Criado Perez, *Invisible Women: Data Bias in a World Designed for Men*, New York: Abrams Press (2019); Bonnie Stabile, Aubrey Grant, Hemant Purohit, and Kelsey Harris, "Sex, Lies, and Stereotypes: Gendered Implications of Fake News for Women in Politics", *Public Integrity*, Vol. 21, Issue 5, 2019, https://doi.org/10.1080/10999922.2019.1626695; Van Der Wilk, A. (2018). Cyber Violence and Hate Speech Online against Women: Women's Rights & Gender Equality: Study for the FEMM Committee. European Parliament. https://dspace.ceid.org.tr/xa twigs raomlui/handle/1/889; Sarah Shoker, *Military-Age Males in Counterinsurgency and Drone Warfare*, London: Palgrave Macmillan (2021); Katharine Millar, James Shires, and Tatiana Tropina, *Gender Approaches to Cybersecurity: Design, Defence and Response*, Geneva, Switzerland: United Nations Institute for Disarmament Research (2021); Marie Segrave and Laura Vitis (eds.), *Gender, Technology and Violence*, New York: Routledge (2017).
33 The Malaysian Centre for Constitutionalism and Human Rights (MCCHR), "Cyberharassment in Malaysia: What Do We See Happening?", 31 January 2018, https://mcchr.org/2018/01/31/cyberharassment-in-malaysia-what-do-we-see-happening. UN Women, "Online Violence".
34 Citron (2014), Filipovic (2007), Franks (2011), and Mantilla (2015) as cited in Sobieraj, p. 4.
35 Sobieraj, p. 10.

36 Citron (2014), Filipovic (2007), Franks (2011), and Mantilla (2015) as cited in Sobieraj, p. 4.
37 Sobieraj, p. 103. See also, Emma A. Jane, "Feminist Flight and Fight Responses to Gendered Cyberhate", in Segrave and Vitis, p. 48.
38 Nina Jankowicz (18 May 2021); Lucina Di Meco, (18 May 2021), and Gabrielle Bardall (18 May 2021) CENS webinar. See also – event report.
39 Sobieraj, p. 103; Emma A. Jane, "Feminist Flight and Fight Responses to Gendered Cyberhate", in Segrave and Vitis, p. 48; Nina Jankowicz, "Malign Creativity: How Gender, Sex, and Lies Are Weaponised against Women Online", Panel 2, 18 May 2021, in "CENS & The High Commission of Canada Webinar Series", pp. 28–32; "Defining the Problem," #Shepersisted, https://www.she-persisted.org/why; "CENS & The High Commission of Canada Webinar Series." See also - event report.
40 Sobieraj, p. 115.
41 Nina Jankowicz, Jillian Hunchak, Alexandra Pavliuc, Celia Davies, Shannon Pierson, and Zoe Kaufmann, "Malign Creativity: How Gender, Sex, and Lies Are Weaponized against Women Online", The Wilson Centre, Science and Technology Innovation Program, January 2021, https:// www.wilsoncenter.org/publication/malign-creativity-how-gender-sex-and-lies-are-weaponized-against-women-online; Lucina Di Meco, "Gendered Disinformation, Fake News, and Women in Politics", Council on Foreign Relations, 6 December 2019, https://www.cfr.org/blog/gendered- disinformation-fake-news-and-women-politics; Bonnie Stabile, Aubrey Grant, Hemant Purohit and Kelsey Harris, "Sex, Lies, and Stereotypes: Gendered Implications of Fake News for Women in Politics", *Public Integrity*, Vol. 21, Issue 5, 2019, https://doi.org/10.1080/10999922.2019.16 26695; Julie Posetti, Nabeelah Shabbir, Diana Maynard, Kalina Bontcheva, and Nermine Aboulex, "The Chilling: Global Trends in Online Violence against Women Journalists", UNESCO, April 2021, https://unesdoc.unesco.org/ark:/48223/pf0000377223
42 Gulizar Haciyakupoglu and Yasmine Wong, "Gender, Security, and Digital Space: Issues, Policies, and the Way Forward", *RSIS Policy Report,* 13 December 2021, https://www.rsis.edu.sg/rsis-publication/cens/gender-security-and-digital-space-issues-policies-and-the-way-forward/#.YrBOjnZByM9. Fitriani Bintang and Sarah Shoker, "CENS and The high Commission of Canada Webinar on 'Gender, Security and Digital Space: Exploring Risks, Opportunities and Security Implications'", RSIS Event Report, 11, 18, 25 May 2021, https://www.rsis.edu.sg/wp-content/uploads/2021/08/Event-Report-for-Gender-Security-and-Digital-Space-1.pdf.
43 Ibid.
44 "Covid-19 Shows We Need More Than Basic Internet Access — We Need Meaningful Connectivity", Alliance for Affordable Internet, 27 May 2020, https://a4ai.org/covid-19-shows-we-need-more-than-basic-internet-access-we-need-meaningful-connectivity/
45 Ibid.
46 Outside of this edited volume, Sobieraj also calls for "a shared language around what it means to be targeted online" as, she believes, this would "help advocates press for improved institutional responses that center victims" (p. 150). The problem with definitions and the need for shared understandings have been a lingering discussion among some scholars working on influence operations, including disinformation campaigns. See Alicia Wanless and James Pamment, "How Do You Define a Problem Like Influence?", *Journal of Information Warfare*, Vol. 2019, Issue 18.3, pp. 1–14, https://carnegieendowment.org/files/2020-How_do_you_define_a_problem_like_influence.pdf/
47 Segrave and Vitis, 2017, pp. 8–11.
48 Ibid, p. 8.

49 Sobieraj, 144; Emma A. Jane, "Feminist Flight and Fight Responses to Gendered Cyberhate", in Marie Segrave and Laura Vitis (eds.) *Gender, Technology and Violence,* New York: Routledge (2017).
50 Cassandra K. Shepherd, "The Role of Women in International Conflict Resolution", *Hamline University's School of Law's Journal of Public Law and Policy,* Vol. 36, Issue 2, Article 1 (2015). Available at: http://digitalcommons.hamline.edu/jplp/vol36/iss2/1
51 Please note that we collated the chapters of this edited volume in mid-2022, and the review process ended in January 2022.
52 Sobieraj, p. 10.
53 James Bridle, *New Dark Age,* London: Verso (2018), p. 13.
54 Ibid.

PART I

2

THE WOMEN, PEACE AND SECURITY AGENDA IN DIGITAL SPACE

Tamara Nair

Introduction

The United Nations (UN) Security Council Resolution 1325: Women, Peace and Security (WPS) agenda (together with subsequent resolutions), created by a conflict but rooted in human security, is suggested here as a possible framework for understanding and addressing issues of gender inequality and violence against women and girls in a digital world. Year 2021 marked the 21st year of the resolution and it seems timely that the agenda itself moves into a new terrain to address gender equality in a new realm of operations. This chapter aims to provide an overview of gendered inequalities in the digital ecosystem by discussing some of the problems and issues in a 'public' space that have for the most part been 'invisibilised'.[1] A feminist lens of ignorance or non-knowledge studies is briefly employed to study how variegated applications of power and ignorance/non-knowledge limit the capacity of groups and individuals to exercise a political agency in the digital realm as is the case in the physical world. In addition, the four pillars of the WPS agenda – prevention, protection, participation and involvement in relief and recovery – provide a new framing through which the security of women and girls can be ensured in the digital space.

The gendered impacts of digital technology, and its connection with peace and security, are a serious policy 'blind spot'. Much like in the physical realm, the idea of women's equality, in general, and their safety and well-being, in particular, has been under-securitised in this new arena for a number of reasons that will be discussed shortly. Gender inequality and gender-based violence in the digital world, as they are in the physical world, are indicators of more widespread social fractures and disruption. Yet, we do not see much action in addressing this 'blind spot' when we look at digital security architecture and the governance of this virtual space. There really is a little movement in addressing this gap in policy

DOI: 10.4324/9781003261605-3

discussions. Much of this has to do with the limited data on women's presence in, and usage of the digital space. How much do we know about women's usage of the digital space and the types of obstruction and/or harassment they face in participating in this space? This is further discussed below.

Digital technologies have only served to intensify tensions between national security and security of individuals, and policies or laws set in place to ensure such security in the digital ecosystem. As a matter of fact, the orientation of digital data is such – devoid of attention to people and places[2] – that it sheds light on the fragility of legal knowledge, which becomes 'increasingly "undone" by digital technologies and future-oriented security practices'.[3] To address these rapid changes, policymakers have opted to explore areas of non-knowledge as they emerge in controversies of mass surveillance, fraud, harassment and the like as they would in the physical realm, through systems of governance that, once again, leave out groups of interest, be it women, sexual minorities or other minority groups. What is required is to advance more critical approaches to security, especially for the protection of women and girls, and to engage with different areas of security studies including assessing available international frameworks that can be used to add a level of buoyancy and longevity to digital security policies. While existing legal frameworks form the backbone of policy formulation, new areas of knowledge can act as the 'skin and bones' upon which to build what will undoubtedly be an organic construct, growing and expanding, keeping up with the digital 'Proteus'. As mentioned at the start of this chapter, one such area of knowledge is the WPS framework.

With the above, I present the chapter's roadmap here: the next section briefly examines women's usage of digital space, marking out particularistic experiences of women that can be constituted as threats to their mental and physical well-being. I will also aim to highlight reasons for the lack of engagement with these gendered insecurities by introducing ideas of ignorance and vulnerability. The subsequent section discusses a WPS framework and its importance in governing the digital space *vis-à-vis* women. The final section concludes the chapter with recommendations on how these insecurities may be addressed using the WPS framework pillars of protection, participation, prevention and the role of women in relief and recovery.

Old threats, new spaces

The WPS agenda, starting with the UN Security Council Resolution 1325, and gaining nine more in subsequent years,[4] is the most highly recognised, significant global framework for addressing 'gender equality in military affairs, conflict resolution and security governance'.[5] The agenda has its provenance in conflict in the 1990s, namely, but not isolated to, the Serbian and Rwandan wars' impact on women and girls. There is massive literature surrounding the agenda over the two decades of its existence, and I will not attempt a description of it here. Rather, I would like to draw attention to, as Laura Shepherd writes, the rather porous borders of the agenda and the extent to which the agenda must change to address new problems[6] that affect already imbalanced gender relations. The 21st year of the agenda marks new and evolving threats to the security of women and girls, in a new space but dragging along with it, old threats to, the treatment of and tirade against one group, which despite forming 50 percent of the world's population is seen as a minority interest in security planning.[7] But given the maturity of the agenda, it now seems poised to take on these new challenges facing women – starting from harassment, exclusion and downright threat to life in the new digital space.

There are as many reports, blogs, articles, protests and even laws against threats or harassment against women online as there are actual incidents of violence faced by women and girls in the digital space. Yet, we do not necessarily see an abatement of these unlawful and dangerous activities despite such justified 'noise' brought up by women's groups or human rights advocates. This is unfortunately evidenced by examples of suicides of prominent female artists as a result of cyber-bullying,[8] cases of 'revenge porn'[9] being spread online by disgruntled former or current intimate partners to death threats and the hate speech,[10] with a particular feminist twist, directed at women who may be exercising their freedom to express opinions on a social media platform. The online space, then, has become an extension of the physical world where inequality and discrimination have been diffused through the technological boundary. We think of technology as being gender neutral but it is in fact highly gendered at its very inception. At a recent *Foreign Policy Virtual Dialogue* in collaboration with *Our Secure Future*, which examined the WPS agenda for the digital age, it was mentioned that the creators of current digital technologies did not create platforms where all individuals would be safe simply because they came from a world where their safety was never threatened in similar ways.[11] The social context within which new technologies are created and later embedded is where misogynistic behaviour resides and, therefore, women as well as girls continue to face old threats, now in new places.

Suppression of one's freedom of speech and expression, 'Cyber-Touch',[12] breach of dignity and violation of privacy; all these constitute Violence Against Women (VAW). Seventy-four percent of countries, although passing or having passed legislation on cybercrimes, lack an adequate mechanism, with concurrent failures on law enforcement, to effectively address the online VAW.[13] Examples abound of such violence in digital space and there are sadly too many to name here. Suffice to say, activists and researchers, for example, the London School of Economics WPS blog,[14] do a good job in bringing forth many examples of such violence. What is of concern here is the non-engagement of policymakers with this area of vulnerability that is classified as 'non-knowledge' as far as dominant security, as well as cybersecurity discourse, is concerned.

Vulnerability as non-knowledge

The idea of 'non-knowledge' or Ignorance Studies is rather new as a theoretical framework but has gained traction in recent years to explain the lack of interest in knowing particular discourses or otherwise, also described as the wilful suppression of particular types of knowledge, especially those around vulnerability in, for example, security assessments and policymaking.[15] Nowhere is its use more applicable than when discussing the exclusion of specific vulnerabilities felt by women. I must stress at this point that although vulnerability and the ignorance of it has been explored in security studies,[16] as well as in gender and feminist studies,[17] it has seemingly not made an appearance in *gender and security studies* as yet.

Epistemologists who study ignorance have actually made a convincing case that ignorance is not a 'mere lack of knowledge but rather actively produced and maintained'.[18] We take from this point of view that social forces undergird the production and reproduction of ignorance as it would for knowledge production of conventional or dominant frameworks of understanding the world. To understand the social construction of ignorance, we must first understand that ignorance is not a consequence of limited knowledge; rather, it is borne of the politics behind differing interests. In their introduction to the special issue on feminist epistemologies in ignorance, Tuana and Sullivan urge feminists to study ignorance more carefully because, as they stress, ignorance is often intertwined with practices of oppression and exclusion.[19] What is more engaging to the feminist scholar is how Erinn Gilson takes their ideas further, combining discourses surrounding ignorance and oppression with what we think of as vulnerability or what it means to be vulnerable. It is from this connection that I see where the ideas above can move from scholarly discussion to policy action.

Gilson argues that the denial of vulnerability is a common phenomenon and it is both ethically and politically dangerous to think so, and she adds that an awareness of vulnerability is central to removing forms of violence and oppressive social relations.[20] Vulnerability is exposed by violence. But with a greater awareness of this vulnerability, one can come up with better political responses or policies to it, which can then stop the trigger-happy reaction to vulnerability with some form of 'capacity' or 'resilience' – words that are often over-used, but also often misunderstood. Such forms of 'problem-solving', or the creation of invulnerability, end with existing violence being handled by other forms of violence or punitive acts that portend to protect individuals, in this case women, from their vulnerability but in fact, do more harm. For example, protecting women from gender-based violence in the public space (including the digital realm) by keeping them in private space (do not engage if you do not want to be targeted) or conducting capacity-building workshops without understanding the social and cultural contexts of women's lives (teaching digital safety without understanding why women are targeted in the first place). Norton's website, in celebrating International Women's Day, has a page for cyber security for women, which outlines the 'dos' and 'don'ts' of engaging in the digital world:

1. *Don't share passwords,*
2. *Don't leave your webcam connected,*
3. *Don't share more than necessary,*
4. *Don't meet online acquaintances alone,*

5. *Reveal only as much as needed,*
6. *Update all operating systems on your devices,*
7. *Secure your devices with anti-virus software,*
8. *Read the fine print,*
9. *There is no such thing as 'freebies' and*
10. *Block people you don't want to interact with.*[21]

As to why the company would think these are specific problems only to women is not clarified. It also asks women to use 'common sense' in engaging in digital space and to 'not feel guilty' in trying to protect themselves from harassment,[22] almost laying blame for harassment with women and their 'feelings'. Such a simplistic understanding does little to alleviate problems of VAW, as they manifest online. These are the types of societal constructs that need to be understood if one is to understand the true meaning of vulnerability, as commonly understood. But there is another aspect to vulnerability that can emerge as a positive in knowledge creation.

Vulnerability as human nature?

From Gilson's work discussed above, I ask the question: if vulnerability is a fundamental part of human nature, as also propounded by feminist philosopher and gender theorist, Judith Butler, why is it not open to greater scrutiny but rather, left to manifest in a singular definition of weakness, dependency, inability and defencelessness?

In this way, vulnerability is not just a negative state but can be a position of power – a way of learning and moving away from a static understanding of helplessness to a dynamic state of engagement to address issues of concern to all such as security in the digital ecosystem. Thinking of vulnerability (specifically here, of women,) as a negative, maintaining this ignorance of vulnerability and not fully understanding the impacts this will have on online VAW illustrate a cultivated ignorance, imported from unresolved/unexamined/unrecognised issues of gender inequality that pervade the social lives of men and women in the physical world. This connection between vulnerability and ignorance is in line with the capitalist socio-economic system, which is often a recurring theme when examining forms of disempowerment, oppression and ideas of resilience.[23] Suffice to say, the ignorance of states of vulnerability – a social construct in and of itself – prevents a clear understanding on how vulnerability can present itself as information that should go into effective policies to protect against the online VAW. This will be examined in greater detail below. The discussion here continues to move on to one particular example of how vulnerability is created and how the ignorance of this vulnerability is not only 'internalised', but also maintained.

There certainly is no impetus at present to indicate any kind of urgent need to veer away from current digital security policies to incorporate a gendered understanding of security in the digital space. This is simply because there is little information that signals this need. Caroline Criado Perez writes that this stems from the notion that what is male is universal. Not only is this a consequence of the gender gap, it is also a cause of it.[24] There is no importance to gather gendered data since women's experiences are seen as a niche area, as Criado Perez puts it, and a subjective point of view. According to (a masculinised) worldview, we know women are vulnerable; that this vulnerability can take many forms, and sometimes, even understood as common knowledge, is self-created. What is male is understood as being representative but what is female, then, is a 'unique' situation, out of the understood and accepted universal spectrum. I interpret her ideas as taking us to heteronormative understandings of how and why expected gender behaviour is manifested, where vulnerability is a weakness, associated with the feminine and opposed to invulnerability and strength – both masculine traits. Therefore, in a masculinised worldview, vulnerability cannot be seen as a strength; so, we always seem to have policies to build *resilience* – a magical word that aims to solve problems plaguing humans undergoing any form of crisis – without truly understanding the nature of the causes of manifestations of vulnerability. This accounts for a limited approach to digital security based on existing frameworks of (masculinised) knowledge (of security), denying the users, in this case, women, the agency to contribute to what and who needs to be included, who needs to be listened to and what is important in decision-making around these security policies.

Framing an issue in a closed and limited fashion stifles alternative approaches, denies access to different areas of engagement and reinforces orthodoxy. If policymakers see specific vulnerabilities as springboards for learning and action rather than viewing them as forms of defencelessness, we would be gathering vital disaggregated data necessary for effective cybersecurity policies aimed at protecting *all* users.

WPS in the digital world – addressing ignorance/ non-knowledge

The WPS agenda puts women at the table of security discussions. Although the initial conception is around military conflict, there is no closed framing that prevents the agenda from being useful in other forms of security including that of human security, humanitarianism and human rights. I have discussed elsewhere the role the agenda can play in involving women in natural disaster-management policies[25] as well as women's economic security.[26] It is not a very far leap to acknowledge the usefulness of the WPS agenda then, in the digital space, as a legitimate framework to include and protect women and girls, and prevent any harm and harassment aimed at them in this platform. In order to extend the agenda's framing into digital security, we need to understand why this gender inequality pervades digital security architecture.

Power/Knowledge

As a concept, numerous scholars and no less than French philosopher Michel Foucault himself have dedicated time to the study of Power.[27] Power is everywhere and it operates at many levels. When you study the gendered impacts of security, politics or economics or almost any other kind of gendered investigation, you are investigating Power – who has it and how is it used? Insufficient political will to break barriers for gender equality in the digital space, limited vision as to what needs to be done to protect and empower women, entrenched cultural views of women that lag behind the realities of their presence and involvement in the digital ecosystem; all these happen because of how Power operates between genders. New knowledge, as in the security threats in the digital space, is never complete and almost always opens up new areas of the unknown. New threats create new situations where it unsettles and unleashes 'fresh debates on how to respond, when to respond, who may respond, who should respond but is incapacitated from doing so'.[28] Under such scenarios, ignorance can be magnified and how we cope with it also sheds light on how not knowing/not wanting to know can be an important resource or mechanism to hold on to Power. If we go by Foucault's thinking, security measures can be seen as a rationality of government in which effective control of the population is achieved not by 'governing more', but by 'governing less', done through mechanisms of self-discipline and self-regulation.[29] These mechanisms, including using specific types of knowledge, are defined as the Power mechanisms that stem from dominant (knowledge) systems that control existing frameworks around particular issues/problems such as cyber security. So, those who control particular knowledge(s) have the power to control what is discussed, how it is discussed and who discusses specific security issues. Therefore, by ignoring other knowledge systems that lie on the peripheries of the dominant discourse, power is maintained within particular specialists, excluding even those who can be most affected by their disciplinary techniques.

Security policy formulation and implementation are undertaken by (male-dominated) national governments, often drawing very heavily on (male-dominated) national agencies using masculinised knowledge – with an overriding emphasis on technology, science, logistics and management that puts aside social factors including that of gender.[30] This also highlights the power of certain types of knowledge over others. It would be untrue to state that women are never invited to participate in discussions on security policy and planning. After all, inviting women in such high-level deliberations makes for great optics in politics. However, they are not invited to redefine the problem or examine closely its causes. Therefore, the female experience in digital space and the actual nature and environment of participatory security deliberations to which they are invited to have very little connections. As Cynthia Enloe writes, 'rarely are they made visible as thinkers and actors'.[31] These knowledge-formation and policy-formation processes are highly problematic at best, and they can be dangerous in cases such as increased online female radicalisation where attention to causes of such behaviour is still not necessarily gendered.[32] People often forget that women are not always the victims of violence; they can be perpetrators too. The WPS agenda tries to tilt this power imbalance through its four pillars of protection, prevention, participation, and relief and recovery.[33] But before I demonstrate that, we need to examine how the idea of vulnerability should be shaped in the minds of policymakers.

Vulnerability as a positive

There certainly is an aura of ignorance surrounding vulnerability, as an uninterrogated background assumption of negativity and weakness – a transient condition that affects only some for a period of time. The nature of vulnerability requires a move from a *de facto* negative understanding to one that is more ambivalent as something being both 'affected and affecting in both positive and negative ways, which can take diverse forms in different social situations …'.[34] Vulnerability should be understood as a primary and fundamental common condition and, in more general terms, encompassing ideas of affectivity and openness to a change and learning, and thus becomes a condition of a potential that makes other conditions possible.[35] It is in this understanding that vulnerability becomes an important agent of change.

The openness around understanding vulnerability can make it easy for it to absorb frameworks, such as the WPS agenda, because it is in this vulnerable condition, people will be most open to want to contribute, to change and to learn, and alter basic structures of 'subjectivity, language and sociality'.[36] That also highlights the importance of the participation of those who are 'vulnerable' in particular circumstances and situations/cases.

It is with this mindset that policymakers should invite women to the table when discussing issues of security, equality and protection in the digital world, with the full appreciation that the digital environment provides one with the necessary tools to succeed and in this day and age, almost everyone is part of it in one way or another. Policymakers should think beyond growth and development and start considering how to make that progress achieved thus far be a more sustainable one. In addition, it would be prudent to (1) understand the different manifestations of threats in contemporary times, (2) move beyond a narrow definition of 'security' and (3) be more inclusive in creating effective policies. Below, I present recommendations on how the WPS framework might facilitate this.

Forging ahead …

From the above discussions, we see a need for policymaking around digital security or even broader security matters to acknowledge areas of non-knowledge or ignorance more generally and that of vulnerability experienced by some. What might be the best way to incorporate gender inequality and gendered security threats in the digital space?

VAW in digital space undermines aims of the WPS agenda. Namely that of sustainable peace and gender equality. As mentioned at the start of the paper, VAW is often the 'canary in the coal mine' for wider implications of unrest and conflict in society, regardless of whether that gender-based violence is in the physical or digital realm. Therefore, as we would employ the agenda's framework in the real-life context, we need to do the same in the digital ecosystem.

Increasingly, sophisticated technologies have created new means of violence against, surveillance of and squashing opposition from women. For example, both men and women are affected by cyberstalking, but a survey in India found that victims aged between 18 and 32 were predominantly female and a research in Argentina shows that a woman's mobile phone is one of the first items to be destroyed by a violent partner.[37] While the WPS agenda has focused on conflict situations, its attention in the digital world is long overdue. Harm in the virtual world can be experienced in reality with severe consequences. Amnesty International's study of online harassment of women across eight countries yielded some rather alarming findings, as detailed in a participant's response:

> I had one incident when I got an email from the FBI; they needed to talk to me about some activity related to my blog. There was a white supremacist who was actively trying to find out where I live. That took it to another level … I had to be very deliberate about my posting for a year after that.[38]

As Arimatsu and Rees comment on their blog on online VAW:

> …: it is incumbent on states to adopt and implement comprehensive gender-sensitive policies to address this growing problem. Non-action would amount to failure on the part of states to comply with their international obligations …[39]

These international obligations would include the Beijing Platform for Action, the UNSCR 1325 and other resolutions under the WPS agenda; Convention on the Elimination of All forms of Discrimination Against Women (CEDAW) and other human rights instruments states might be a party too.

The operationalising of the WPS agenda has seen a mixed bag of positive and not so positive responses, as it is realised in its conventional understanding. At the international level, there has been many rhetorical promises by various international agencies, not in the least the Security Council itself *viz*, the agenda; its implementation in international security and peace-building and concrete action includes the promoting of greater gender balance in the UN military and police contingents.[40] One particular way of assessing levels of acceptance and 'success' has been the adoption of National Action Plans (NAPs) by countries. But this too has shown the width of interpretation of the agenda rather than some form of universal 'acceptance' in terms of women's role in international security.[41] Despite these setbacks, the future of the agenda will be manifold especially as we realise how our security threats are being redefined.

Below are some recommended actions along the WPS framework pillars as a starting point. Before expanding on these recommendations, it is important to note that in order for these actions to be considered/pursued, there needs to be a fundamental shift towards openness, inquisitively – a desire to want to know; a move away from a unitary Power/knowledge base to a broader base of knowledge(s), including incorporating non-knowledge or ignorance in areas of importance. With that in mind, I present four recommendations below.

Recommended actions

WPS Pillar: Prevention

1. Reassess digital security policies with the explicit intention to include 'vulnerability' as an area of engagement in preventing the online VAW.

   It would be in the best interest of both public and private security entities to relook at their current digital security policies for gaps in understanding how vulnerability manifests for different groups of people. In identifying these gaps, we will be better equipped to realise the shortcoming of existing methods of preventing forms of harassment and abuse against women. We can then move on to creating more targeted response to the online VAW and also identifying groups that need to be consulted in policy making.

WPS Pillar: Participation

2. Redefine/widen definitions of security, knowledge and vulnerability with the explicit intention to include greater diversity and, in particular, women's voices, in digital security policies.

   Effective and sustainable policies are the ones that incorporate consultations with a broad base of diverse participants and security policies should be no different. By widening definitions, we realise that generic digital security policies cannot apply to all users if the intention is to keep everyone safe and able to exercise their right to participate in this public space. For example, as mentioned above, the creation of digital technologies and the governing as well as the securitising of digital space has been a very gendered affair – omitting women and non-binary groups. Hence, this particular pillar focuses on including women to, not only join discussion on digital security, but also be able to direct conversations and shape the agenda as well.

WPS Pillar: Protection

3. Collect disaggregated data on harassment/violence against different groups of users of the digital space.

   This will require an openness to accepting different forms of knowledge production as important in security dialogue. With this comes the appreciation of having diversity in data collection and analysis and the different types of knowledge that will drive this endeavour. It is only through a broad capture of data and information can effective, evidence-based policies be formulated. These are exactly such policies that are required for a greater protection from VAW, both in the virtual and physical worlds.

WPS Pillar: Relief and Recovery

4. Focus on women's participation and feedback/knowledge especially in rebuilding efforts after crisis situations such as a global pandemic or a financial downturn.

   After a crisis, the general pattern is to usually fall back to known and comfortable ways of acting and doing and letting a crisis 'go to waste' as an impetus for change for the better. The relief and recovery pillar is often overlooked given the importance of preventing VAW and continuously protecting women from it. But it is this pillar that provides the most relevant information pertaining to harassment and abuse in both the digital and physical spheres because it speaks out of lived experiences, especially in the aftermath or at the tail end of an upheaval. Here is where all the other pillars can be strengthened, by the inclusion of women in policies aimed at rebuilding and restoring lives. Lesson learnt from the victims of VAW should reflect strongly in digital security policies moving forward.

   As a final note, I would like to add that the powers we hope to shatter are long standing and deeply entrenched in our cultures, societies and at all levels of governance – from local, national and international. It is not a new struggle and it is unlikely to end any time soon. But in allowing it to fester, we help perpetuate (and propagate) violence against half the world's population. This violence against women and girls is increasingly taking hold in the digital sphere and in time will engulf it if nothing is done to stop it in its tracks. The safety and security of any user in the digital space should not be compromised because that would simply lead to the security of all users being compromised in time to come. What is needed is a conscious effort by policymakers, at all levels of governance, to be cognizant of that fact and act in a manner that upholds the rights and dignity of all.

Notes

1 To be invisibilised means to be (in some cases, intentionally) unacknowledged or to be side-lined. This has to do with the understanding of the importance of certain groups' decision-making based on whether they hold necessary knowledge/information that can be advantageous or add to existing narratives surrounding particular issues.
2 Aradau, Claudia. "Assembling (Non)Knowledge: Security, Law, and Surveillance in a Digital World". *International Political Sociology*, 11 (2017): pp. 327–342.
3 Ibid: p. 329
4 Resolution 1325 (2000) introduces the WPS agenda, while the remaining resolutions modify/add on to it. These are: 1820 (2008), 1888 (2008), 1889 (2009), 1960 (2010), 2106 (2013), 2122 (2013), 2242 (2015), 2467 (2019) and 2493 (2019). Together they

form the international policy framework on the WPS and were adopted in response to persistent advocacy from civil society. The obligations in the resolutions extend from the international level to the national level.

5 Basu, Soumita, Paul Kirby and Laura J. Shepherd. "Women, Peace and Security: A Critical Cartography", (pp. 1–25). In Soumita Basu, Paul Kirby and Laura J. Shepherd (Eds.), *New Directions in Women, Peace and Security*. (Bristol, UK: Bristol University Press), 2020: p. 1.

6 Shepherd, Laura, J. "Knowing Women, Peace and Security: New Issues and New Modes of Encounter". *International Feminist Journal of Politics*, 22, 5(2020): pp. 625–628.

7 Post, Abigail S. and Paromita Sen. "Why Can't a Woman Be More like a Man? Female Leaders in Crisis Bargaining". *International Interactions*, 46, 1(2020): pp. 1–27; Lawless, Jennifer, L. "Women, War, and Winning Elections: Gender Stereotyping in the Post-September 11th Era". *Political Research Quarterly*, 57, 3(2004): pp. 479–490, and Koch, Michael, T. and Sarah A. Fulton. "In the Defense of Women: Gender, Office Holding, and National Security Policy in Established Democracies". *The Journal of Politics*, 73, 1(2011): pp. 1–16.

8 Marks, Hayden. "Cyberbullying and the Tragedy of Hana Kimura". *The Diplomat*. 05 June 2020. https://thediplomat.com/2020/06/cyberbullying-and-the-tragedy-of-hana-kimura/; Zuppello, Suzanne. "Why We Let Famous Women Get Bullied Online". *Rolling Stone*. 21 July 2016. https://www.rollingstone.com/culture/culture-features/why-we-let-famous-women-get-bullied-online-93244/; Kim, S., Kimber, M., Boyle, M.H., and Georgiades, K. "Sex Differences in the Association Between Cyberbullying Victimization and Mental Health, Substance Use, and Suicidal Ideation in Adolescents". *The Canadian Journal of Psychiatry*, 64, 2(2019): pp. 126–135; among numerous other sources that indicate cyber bullying is prevalent against female celebrities and the females present a significantly higher number of suicide victims because of cyber bullying.

9 Mckinlay, Tahlee, and Tiffany Lavis. "Why Did She Send it in the First Place? Victim Blame in the Context of 'Revenge Porn'." *Psychiatry, Psychology and Law*, 27, 3(2020): pp. 386–396; Dodge, Alexa. "Trading Nudes like Hockey Cards: Exploring the Diversity of 'Revenge Porn' Cases Responded to in Law". *Social & Legal Studies*, 30, 3(2021): pp. 448–468; Bloom, Sarah. "No Vengeance for Revenge Porn Victims: Unraveling Why This Latest Female-Centric, Intimate-Partner Offense Is Still Legal, and Why We Should Criminalize it". *Fordham Urban Law Journal*, 42 (2014): p. 233, among many others.

10 Edstrom, Maria. "The Trolls Disappear in the Light: Swedish Experiences of Mediated Sexualised Hate Speech in the Aftermath of Behring Breivik". *International Journal for Crime, Justice and Social Democracy*, 5, 2(2016): pp. 96–106; Scott, Jennifer. "Misogyny: Why Is it not a Hate Crime?" BBC News. 15 March 2021. https://www.bbc.com/news/uk-politics-56399862; Piscopo, Jennifer. "Being a Woman in Politics Shouldn't Come with Death Threats". *Ms*. 12 February 2020. https://msmagazine.com/2020/12/02/violence-against-women-being-a-woman-in-politics-shouldnt-come-with-death-threats/

11 Swisher, Kara. (Panellist) Foreign Policy Virtual Dialogue (in collaboration with Our Secure Future): *WPS for the Digital Age: Putting Gender on the Tech Agenda*. 06 May 2021. https://oursecurefuture.org/news/virtual-dialogue-women-peace-security-digital-age

12 This is a term coined by the UN broadband commission for digital development. Refer to: UN Broadband Commission. "'Cyber Violence Against Women and Girls' A worldwide Wake-Up Call". *Discussion Paper*, UN Broadband Commission for Digital Development Working Group on Broadband and Gender. UN Broadband Commission, 2015. https://www.broadbandcommission.org/Documents/reports/bb-wg-gender-discussionpaper2015-executive-summary.pdf

13 Gurumurthy, Anita and Amrita Vasudevan. "Equality, Dignity and Privacy Are Cornerstone Principles to Tackle Online VAW". *LSE Women, Peace and Security*

Blog, 04 December 2017. https://blogs.lse.ac.uk/wps/2017/12/04/equality-dignity-and-privacy-are-cornerstone-principles-to-tackle-online-vaw/

14 For more on blogs from the WPS Centre in LSE, please refer to: https://blogs.lse.ac.uk/wps/?s=WPS+and+digital+security

15 Jackson, Richard. "The Epistemological Crisis of Counterterrorism". *Critical Studies on Terrorism*, 8, 1(2015): pp. 33–54; Roberts, Joanne. "Organizational Ignorance" (pp. 361–369). In Mathias Gross and Linsey McGoey (Eds.), *Routledge International Handbook of Ignorance Studies*. (London, New York: Routledge), 2015.

16 Aradau, Claudia. "Assembling (Non)Knowledge: Security, Law, and Surveillance in a Digital World". *International Political Sociology*, 1, 4(2017): pp. 327–342; Stavrianakis, Anna. "Requiem for Risk: Non-knowledge and Domination in the Governance of Weapons Circulation". *International Political Sociology*, 14, 3(2020): pp. 233–251, among several other works.

17 Scholars Nancy Tuana and Shannon Sullivan have edited an entire special issue on this topic in the journal, *Hypatia* (21, 3(2006)). This issue is worth exploring, especially for those interesting in feminist epistemologies of ignorance. Another interesting work on the misinformation on oppression and obedience of religious women and their assumed negative ideas on feminism can be found in Đonlagić, Halida. "The Effects of the Dominant Public Discourse and the Influence of (Non) Knowledge as a Sign of Resistance/Support to Women's Faith-Based Peace Activism in Bosnia and Herzegovina". *Occasional Papers on Religion in Eastern Europe*, 41, 3(2021): p. 2.

18 Gilson, Erinn. "Vulnerability, Ignorance and Oppression". *Hypatia*, 26, 2(Spring 2011): p. 309.

19 Tuana, Nancy and Sullivan, Shannon. "Feminist Epistemologies of Ignorance". *Hypatia*, 21, 3(2006): pp. i–iii.

20 Gilson. "Vulnerability, Ignorance and Oppression".

21 Norton LifeLock. "Cyber Security for Women". (2019–2021). https://us.norton.com/internetsecurity-privacy-cyber-safety-for-women.html

22 Ibid.

23 Mavelli, Luca. "Resilience beyond Neoliberalism? Mystique of Complexity, Financial Crises, and the Reproduction of Neoliberal Life". *Resilience: International Policies, Practices and Discourses*, 7, 3(2019): pp. 224–239. Capitalist socio-economic systems here is understood as a rationality of government performed through regimes of subjectification that extends the power of the market to all spheres of human activity (see Mavelli, *Resilience beyond Neoliberalism?* above). Those that subscribe to this line of thinking champion resilience by reinforcing and normalising the idea that individuals are ultimately responsible for their social and economic security. (see: Joseph, Jonathan. "Resilience as Embedded Neoliberalism: A Governmentality Approach". *Resilience: International Policies, Practices and Discourses*, 1, 1(2013): pp. 38–52.)

24 Criado Perez, Caroline. *Invisible Women: Data Bias in a World Designed for Men*. (New York, USA: Abrams Press), 2019.

25 Nair, Tamara. "Upscaling Disaster Resilience in Southeast Asia — Engaging Women Through the WPS Agenda". *RSIS Policy Report*. 14 March 2018; Nair, Tamara. "Gender, Humanitarian Emergencies and Security", (pp. 176–195). In Chantal de Jonge Oudraat and Michael Brown (Eds.), *Gender and Security Agenda: Strategies for the 21st Century*. (US: Routledge), 2020.

26 Nair, Tamara. "Working Women and Economic Security in Southeast Asia". *RSIS Policy Report*. 14 November 2019; Nair, Tamara. "Gender and Economic Security in Southeast Asia". *Asian Journal of Comparative Politics Asian Journal of Comparative Politics*, 7, 1(2022): pp. 29–44.

27 The greater or overarching concept of Power, as a disciplinary technique, used by scholars such as Foucault is capitalised in this chapter. However, discussions on 'power', 'powerplays' or 'the powerful' will retain small lettering to distinguish between Power as a concept and power as understand in everyday usage.

28 Gross, Mathias and Linsey McGoey. "Introduction", (pp. 1–14). In Mathias Gross and Linsey McGoey (Eds.), *Routledge International Handbook of Ignorance Studies*. (London, New York: Routledge), 2015.
29 Foucault, Michel, Arnold I. Davidson and Graham Burchell. *The Birth of Biopolitics: Lectures at the Collège de France, 1978–1979*. (Springer), 2008.
30 Woodward, Rachel and Claire Duncanson. "An Introduction to Gender and the Military", (pp. 1–20). In Woodward Rachel Woodward and Claire Duncanson (Eds.), *The Palgrave International Handbook of Gender and the Military*. (London: Palgrave Macmillan), 2017; Basu, Soumita and Akhila Nagar. "Women, Peace and Security", (pp. 212–221). In Tarja Väyrynen, Swati Parashar, Élise Féron and Catia Cecilia Confortini (Eds.), *Routledge Handbook of Feminist Peace Research*. (London: Routledge), 2021.
31 Enloe, Cynthia. *Bananas, Beaches and Bases: Making Feminist Sense of International Politics 2nd edition*. (Berkeley, Los Angeles, London: University of California Press), 2014: p. 34.
32 Hanifah, Hana. "Gendering CVE in Indonesia", pp. 177–194. In Greg Barton G., Mateo Vergani and Yenny Wahid (Eds.), *Countering Violent and Hateful Extremism in Indonesia. New Security Challenges*. (Singapore: Palgrave Macmillan), 2022. https://doi.org/10.1007/978-981-16-2032-4_8
33 For those interested in understanding these pillars and the agenda in greater detail, I recommend perusing the excellent works presented by WPS and feminist scholars in the *Oxford Handbook on Women, Peace and Security* (Oxford, New York: Oxford University Press), 2019 edited by Sara E. Davies and Jacqui True.
34 Gilson. "Vulnerability, Ignorance and Oppression".
35 Ibid.
36 Ibid.
37 The Hindu. "From Lahore to Lucknow, Crimes against Women Spur more Surveillance". 09 April 2021. https://www.thehindu.com/sci-tech/technology/crimes-against-women-spur-more-surveillance/article34281591.ece, and Association for Progressive Communication (n.d.). "How Technology is Being Used to Perpetrate Violence Against Women – And to Fight it". 18 November, 2010. https://www.apc.org/en/pubs/research/how-technology-being-used-perpetrate-violence-agai
38 Amnesty International. "Amnesty Reveals Alarming Impact of Online Abuse against Women". 20 November 2017. https://www.amnesty.org/en/latest/press-release/2017/11/amnesty-reveals-alarming-impact-of-online-abuse-against-women/
39 Arimatsu, Louise and Madeleine Rees. "Why Addressing Online Violence Against Women Matters to the WPS Agenda". *LSE Women, Peace and Security Blog*. 27 November 2017. https://blogs.lse.ac.uk/wps/2017/11/27/why-addressing-online-violence-against-women-matters-to-the-wps-agenda/
40 Oudraat, Chantal and Michael E. Brown. "Gender and Security: Framing the Agenda", (pp. 1–27). In Chantal Oudraat and Michael E. Brown (Eds.), *The Gender and Security Agenda: Strategies for the 21st Century*. (London and New York: Routledge), 2020.
41 Ibid.

3

WHAT CAN INTERNET SHUTDOWNS TELL US ABOUT GENDER AND INTERNATIONAL SECURITY?

Sarah Shoker

Introduction: Gender in digital ICTs and international security

In December 2019, the United Nations (UN) General Assembly hosted its first Open-Ended Working Group on Digital ICTs in the Context of International Security (hereafter, OEWG).[1] At the OEWG, Several UN member states highlighted critical infrastructure protection as central to their national cybersecurity strategies. Digital ICTs are fundamental to sustaining contemporary state–citizen relations, but also introduce technical vulnerabilities that cause policymakers to view critical infrastructure as an ecosystem "of potential future disaster and [...] complex landscape of response."[2] Indeed, the Canadian National Strategy for Critical Infrastructure defines critical infrastructure as the "physical and information technology that facilitates networks, services, and assets, which, if disrupted or destroyed, would have a serious impact on the health, safety, security or economic well-being of Canadians or the effective functioning of governments in Canada."[3]

In practice, the protection of critical infrastructure has generally meant that security practitioners try to identify technical and computational vulnerabilities in the ICT ecosystem. Critical infrastructure's technological foundation is crucially important, but an analysis that does not examine the link between human security[4] and technological robustness can risk obfuscating why critical infrastructure protection is important in the first place. If critical infrastructure protection is necessary for social well-being, then ICT failures will also have negative social consequences. Critical infrastructure is responsible for providing access to energy, health services, and, of course, information technology. When critical infrastructure fails, the social consequences can include the severing of connection between people and essential services like banks, hospitals, and

DOI: 10.4324/9781003261605-4

schools. Infrastructure becomes *critical* infrastructure when its proper functioning is vital to societal well-being; its failure creates existential friction between the state and the people residing within its borders.

This chapter is an introductory assessment on the gendered impacts of internet shutdowns. I use country examples and interviewee experiences to make women visible in a policy domain where gender has often been excluded. While critical infrastructure is usually viewed as a site for malicious interference, less attention has been paid to states who intentionally jeopardize their own digital assets and the gendered consequences that result from this decision. A gender analysis "is a critical examination of how differences in gender roles, activities, needs, opportunities and rights/entitlements affect men, women, girls, boys, non-binary or gender-fluid persons in certain situations or context."[5] Though this chapter's focus is on women and girls,[6] the term *gender* is not synonymous for *women*. Rather, gender is an identifier that impacts all people. In Canada, Gender-based Analysis Plus (GBA+) is applied to other groups, including men, members from the LGBTQIA group, persons with disabilities, Indigenous groups, and racial minorities. To support this analysis, I draw on unique, anonymized interview data conducted prior to the COVID-19 pandemic with persons from African and Caribbean countries who have experienced internet shutdowns or who work on internet governance.[7]

At the OEWG, several civil society members called on member states to conceptualize critical infrastructure protection using a human-centered lens, though "human-centered" is still a term that needs to be explored at these meetings. Though syntactically similar, the term "human security" was rarely mentioned. Moreover, scholars have noted that security approaches that speak broadly of "humans" can obfuscate the unique security needs of women.[8] Other scholars have argued that human security that takes gender seriously is a challenge, rather than a complement, to state-centric security.[9] Whether "human-centered" was used synonymously to mean human security remains unclear.[10] Most civil society representatives at the OEWG appeared to use "human-centered" to distinguish their aims against traditional state-centric conceptions of security. Civil society actors did not shy away from mentioning gender in their formal statements, perhaps indicating that gender-specific cyber vulnerabilities were to be included in "human-centered" visions of security.

Despite this conceptual confusion, the message from civil society actors was clear: instead of viewing critical infrastructure as a purely technical system, discussions on ICTs and critical infrastructure need to make ordinary people visible. Internet shutdowns are especially poignant to this discussion because they are critical infrastructure failures that are usually instigated by governments against their own people. Internet shutdowns also highlight the tension and the dependencies between human rights and state capacity. If, as the Tallinn Manual on international cyber security law indicates, the internet is an "enabler of rights,"[11] it remains that individuals and communities are often dependent on the state for securing rights that are often thought to be inalienable, universal, and at jeopardy from the state itself. People often depend on the state to eliminate vulnerabilities that can undermine human security, even while policymakers acknowledge that states can be just as likely as foreign adversaries to harm their own people.

Women and girls are often underrepresented in international security, which can result in serious negative consequences for global cooperation and stability. In cases where conflict resolutions processes involved civil partners, the risk of conflict occurring again dropped by 64 percent.[12] When women were involved in conflict resolution processes "the chance of achieving a more durable peace rose to 78 percent."[13] Some research even finds that gender inequality within states is a better predictor of international state aggression than variables like democracy and wealth.[14] Indeed, as part of its commitment to Women, Peace, and Security (WPS), the Government of Canada notes that *intra*-state conflict is linked to *inter*-state conflict.[15] Women's meaningful involvement in diplomatic, humanitarian, and security is linked to improved outcomes for state-building, conflict prevention, and resolution.[16] International fora should therefore incorporate gender analyses into discussions about international cybersecurity (which includes infrastructure protection) even if, traditionally, these spaces have struggled to do so—the consequence is improved security for everyone

The rest of this chapter is dedicated to exploring the different types of Internet shutdowns in use today, the gendered setbacks that they enable, and ends by explaining why governments opt to use internet shutdowns. Despite increased global coordination aimed at correcting women's underrepresentation and influence in global cyber discussions, internet shutdowns remain a policy choice where women's rights are often ignored or undermined in the quest for national security—possibly because internet shutdowns are state-initiated rather than understood as a form of cybercrime. Though OEWG civil society actors encouraged states to broaden their security paradigm to include "human-centered" concerns, internet shutdowns illustrate that traditional security definitions persist, to the point where citizens' access to "rights-enabling technology" is viewed as a potential risk to governments.

What are internet shutdowns?

For the purpose of this chapter, I define internet shutdowns as critical infrastructure failures that are initiated by states; they are intentional, voluntary, and used to block digital access within turbulent political environments. They are especially appealing for governments seeking to act quickly "and [that] might have limited capacity for other mechanisms of online control."[17] Though Freedom House estimates that approximately 46 percent of the world's population has experienced "politically-motivated internet shutdowns as recently as 2019," the character of these shutdowns is not identical across cases.[18]

Internet shutdowns can affect fixed-line or mobile internet for an entire country, such as the *complete* shutdown imposed by Egypt in 2008 in response to Arab

Spring protests. Internet shutdowns can be *localized*, such as India's decision to block internet access in Kashmir. There are also *partial* internet shutdowns, which can block certain websites or target specific social media platforms—WhatsApp was the social media platform most frequently blocked in 2020.[19] According to data on social media blockages published by Surfshark, a virtual private network (VPN) provider, 30 out of 54 countries in Africa have blocked access to social media while 28 out of 48 countries in Asia have blocked access since 2015.[20] Fewer social media blockages were recorded in the Americas, with four out of 30 countries instituting restrictions, and in eastern Europe, where four out of 46 countries also blocked access.[21] To date, countries in North America, western Europe, and Australia and Oceania have not instituted social media blockages (*ibid*). *Curfew* internet shutdowns, like those used by Myanmar shortly after the 2021 parliamentary elections, are scheduled and systematic cuts to the internet between certain time periods.[22]

More recent attention has been directed at *internet shutdowns in active conflict zones*, such as when Israeli Defense Forces targeted the Gaza Strip's telecommunications infrastructure with airstrikes in May 2021, causing Palestinians to lose access to internet service and which prevented both civil society actors and journalists from doing their work.[23] Other authors have theorized the existence of *slow* internet shutdowns, which occur when governments use regulatory creep to prohibit, interrupt, or make it too costly to participate or create content online.[24] The popularization of social media taxes in African countries is especially noted. For example, Tanzania's "social media tax" forces bloggers and online content creators to pay the equivalent of 930 USD annually in a country whose GDP per capita is 879 USD.[25] Uganda also introduced a social media tax designed to quell online "gossip" on content outside traditional media, resulting in a drop of at least 2.5 million users three months after the regulation was introduced.[26] In effect, slow shutdowns are "politically ambiguous" because instead of displaying "push-button autocracy" that abruptly disrupts internet or social media access, states use internet policy to "inhibit, preempt, or criminalize certain kinds of online content creation" with the effect being the long-term and pervasive inaccessibility of internet services by ordinary people.[27] Though internet shutdowns are mostly used by low- and middle-income countries, there are notable examples of high-income liberal democracies—such as public agencies in the United Kingdom and the United States—cutting access to internet services, a topic that is further explored below.

Most authors note that internet shutdowns were popularized by the Arab Spring during 2011, when Egypt cut internet access in an attempt to dampen growing civil protests against Hosni Mubarak's 30-year presidency.[28] Since 2011, internet shutdowns have proliferated. In 2019 and 2020, Access Now, a prominent NGO that reports on digital rights, recorded 213 and 155 shutdowns worldwide, respectively.[29] In the first five months of 2021, Access Now counted at least 50 shutdowns across 21 countries.[30] While the frequency and breadth of internet shutdowns worldwide make a country-by-country analysis impossible for the scope of this chapter, there are key themes that cut across country differences. Research conducted by Deborah Brown and Allison Pytlak finds that gendered impacts under internet shutdowns can be summarized into five main themes: personal safety, professional/economic impact, emotional well-being, education, and finding alternative connectivity.[31] While internet shutdowns are proliferating worldwide, the research examining their gendered consequences remains relatively small. Consequently, this chapter's emphasis focuses on assessing the problem with the understanding that greater analysis is needed. This is especially the case because many countries still do not collect sex-disaggregated data,[32] which makes understanding the gendered consequences of internet shutdowns even more difficult to decipher.

Internet shutdowns, gender, and the digital divide

Internet shutdowns are the part of a larger conversation on internet accessibility and online user-generated content. Globally, women are a minority of internet users.[33] The gendered digital divide, the term given to describe the gap between women and men's meaningful access to and use of digital ICTs, is an important predictor of gender equality within states. When women have greater access to digital ICTs—or said otherwise, when the gap between men and women's digital access closes—then the state in question is more likely to experience higher levels of gender equality. This pattern is maintained across low-, middle-, and high-income states.[34] Conversely, the negative effects of internet shutdowns are likely to be exacerbated when they intersect with other social indicators that are relevant to women's full political membership. As explored below, these social indicators include geographic location, income level, the rural-urban divide, the strength of the country's broadband infrastructure, and whether women belong to other marginalized groups.

Extensive research finds that women and girls experience critical infrastructure emergencies differently than boys and men.[35] Internet shutdowns create the conditions for the backsliding of women's equality. Despite this well-known research, few states at the OEWG chose to prioritize gender in their statements to civil society groups. One civil society representative stated that states "spoke about protecting themselves, not protecting citizens or leveraging citizen potential … we barely had any conversation about women even when we spoke about critical infrastructure. Even access to critical infrastructure is differentiated between women and men."[36]

Women's access to the internet fluctuates depending on geographic location, income level, and whether they live in urban or rural environments. Women comprise 22.6 percent of users in Africa and 41.3 percent in Asia.[37] According to one survey conducted across Columbia, Ghana, Uganda, and Indonesia (the country with the highest VPN usage rate due to state censorship of online content), the gender digital divide is closing around basic connectivity. However, women continue to experience "second-rate internet," meaning that they do not have access to "minimum thresholds for regular access, an appropriate device, enough data and a fast connection."[38] Women are also less likely to create online content, especially political and social commentary,[39] and incur disproportionate risks when they do engage on these topics.[40] The exception to the gendered digital divide occurs in high-income states, where women sometimes outpace men in internet use-rates.[41]

When governments initiate internet shutdowns, they are more likely to target mobile internet connectivity rather than fixed broadband lines.[42] Mobile internet connectivity is targeted because in many low- and middle-income countries, broadband infrastructure remains underdeveloped and therefore underused.[43] Effectively, this means that cheaper access to the internet and mobile phones is crucial for narrowing the digital divide, especially since women are more likely to live in poverty and therefore less likely to afford digital services.[44] In Myanmar, for example, the widespread expansion of mobile internet data is largely credited to cheaper mobile SIM cards, from 1500 USD in 2010 to less than 1 USD in 2016.[45] In response, some civil society actors offer technologies that are designed to help users circumvent internet shutdowns by targeting mobile phone users.[46]

Women are disproportionately represented among those living in extreme poverty, both globally and in Uganda specifically,[47] meaning that they have fewer alternatives for accessing online information when governments suspend social media websites. In the case of "slow" internet shutdowns, evidence suggests that women are disproportionately affected by social media taxation. For the approximately one-third of Ugandans who live in extreme poverty (defined as living on 1.90 USD or less per day), digital access is often restricted to social media applications like WhatsApp and Facebook. Without the ability to pay the social media tax, their access to the internet is completely eliminated.[48]

Internet shutdowns can also amount to a form of digital redlining[49] when localized against particular communities, a problem that was amplified by the COVID-19 crisis when much of the world's work and educational institutions were forced to "go digital."[50] This means that the loss of work, wages, and schooling under internet shutdowns can exacerbate social inequality on the basis of identifiers that intersect with gender, like ethnicity and religion. In India-controlled Kashmir, one of the most frequently targeted regions in the world, women's rights activists struggled to provide counseling to those who were targeted with domestic violence due to the communication blockade.[51] This has been especially problematic under COVID-19, since the pandemic has been accompanied by increased rates of domestic violence against women in several countries around the world.[52] Doctors, too, were left without sufficient medical information to treat patients, thus amplifying health disparities based on region, gender, and religion.[53] Without internet access, individuals were unable to obtain information about the pandemic or access social support networks.[54]

Beyond accessing educational and NGO services, internet shutdowns are also highly correlated with reports of government-instigated human rights abuses. There does appear to be a relationship between social unrest and ICT blockages during elections and government transitions, and it is during this period that women are more likely to be targeted with sexual violence and other forms of political disenfranchisement.[55] The following sections contain statements made by interviewees; interviewees have been assigned a number to protect their anonymity.

Interviewee 2 participated in the 2019 Sudanese sit-in and summarized these gendered patterns of violence succinctly: "they rape the women and kill the men."[56] In political climates where physical safety was not a guarantee, women used the internet to verify whether they could leave the house. Additionally, the internet was used to determine whether public spaces were sufficiently secure for their participation in civil protest. The same interviewee stated that the internet shutdown impacted girls and women more than their male counterparts. (The June 3rd sit-in was disrupted by the Transitional Military Council and was later dubbed the "Khartoum massacre" after more than 100 people were killed.) According to Interviewee 2, women and girls "are less likely to be allowed by their parents, guardians, or even husbands to leave the house. So, we depend more on the internet to connect us to the news and to mobilize."[57] The removal of internet access meant that women were more likely to be politically isolated and disconnected from the public sphere. In general, internet access serves important information functions by allowing women to navigate spaces that come with higher risks of sexual violence. "If there are guns outside your house, at least you have the internet and there are people telling you something is happening here and there, at least you don't feel alone."[58]

Why do governments use internet shutdowns?

To speak about internet shutdowns is to also speak about the contentious topic of state control over the creation and use of online information. As mentioned earlier, internet shutdowns are mostly, though not exclusively, used by low- and middle-income countries. Some authors argue that internet shutdowns may be a response to inadequate regulatory capacity. According to De Gregoria and Stremlau, low-income states have limited ability to regulate the circulation of online content, with many having "not developed a systematic way of engaging with [social media companies], including notifying them when content violates national laws."[59] When social media companies do not address complaints surrounding the spread of misinformation, much of which is designed to incite hostility and violence, then internet shutdowns become one solution from a restricted range of policy options.[60]

No country maintains absolute control over the freedom of expression. Importantly, liberal democracies have also struggled to keep extremist content offline. For example, the mass shooter who targeted a mosque in Christchurch, New Zealand, also livestreamed the attack for 29 minutes. The video was watched by hundreds of people on Facebook before the video was pulled off the platform.[61] In response, Canada's minister for public safety at the time warned that if social media platforms could not adequately regulate their online content, then they "should expect public regulation."[62] In general, though, high-income countries are in a better position to make online platforms comply with their regulations. In my own interviews, respondents acknowledged that low- and middle-income states resorted to internet shutdowns because they lacked alternatives for responding to misinformation. As interviewee 3 stated, "you can sue a newspaper in a local court, but you can't sue Facebook … In the traditional world, you can revoke their license, but that can't happen with Facebook or WhatsApp."[63]

Though internet blackouts are mostly used by low- and middle-income countries, there are notable exceptions. In 2019, the British Transport Police blocked wi-fi access in the London Underground to disrupt a protest by an environmentalist activist group called Extinction Rebellion, a tactic that they did not make public prior to the shutdown.[64] In 2011, the San Francisco Bay Area Rapid Transit System (BART) blocked mobile service because protestors were planning to block train platforms in response to BART police shooting and killing an unarmed passenger. According to one source, this was the first internet disruption in US history that was instigated by a government agency.[65] "Without the ability to coordinate their efforts via cell phones, protesters were unable to commit acts of civil disobedience."[66] In both cases, transit representatives cited safety and the protection of their passengers.[67] Governments do not always publicly explain why they choose internet shutdowns, but "safety" during protests is one of the most frequently cited reasons for disrupting access across all countries, though Access Now is quick to note that these justifications are often "excuses" used against peaceful protestors.[68] In an international environment where liberal democracies are thought to act as "norm entrepreneurs" that set the international agenda, liberal democracies risk diluting their own diplomatic influence if they use the same techniques as autocratic states.

> Even my country, which is fairly liberal, sees that if Hong Kong can do it, if India is doing it—which are fairly well-developed countries—then why can't we do it in our own country? Why not shutdown WhatsApp, because what use is it serving other than spreading political misinformation? And my country is not a military dictatorship. But if you see another country doing it, then it becomes the norm. I'm not aware of any country that has been sanctioned over an internet shutdown, are you?[69]

Yet, internet shutdowns are not only a response to a regulatory gap. The fact that police in liberal democracies have cut internet access indicates that shutdowns can be appealing to powerful states. States that initiate internet shutdowns are also likely to use service disruptions in tandem with punitive police and military methods against their own citizens, as in Sudan, where a peaceful protest was met with the military violence.[70] Increasingly, states also use digital platforms to spread misinformation that inflame partisan violence, as in Myanmar, where digital misinformation was used to perpetuate genocidal violence against the Rohingya minority, including mass sexual assault against women and men.[71] Social media companies are consequently trapped within an awkward security dynamic: the removal of content at the request of autocratic regimes facilitates a political environment where citizens are more easily targeted by their governments.

Conclusion: Where do we go from here?

Digital inequality amplifies social inequality. Many of the activities that are indicators of women's equality—like educational attainment, political participation, and social mobility—are increasingly accessed through digital platforms. If ICTs are rights-enabling tools, then their abrupt withdrawal, as in the case of internet shutdowns, not only challenges international stability but installs numerous obstacles for women's full membership into their communities. However, these obstacles have not stopped women from resisting, politically mobilizing, and taking leadership on this issue.[72] On the topic of the Sudanese sit-in, Interviewee 1 notes:

> What was surprising for the public was that women leaders were the ones who were talking about internet shutdowns, more so than male leaders. Women were more in need of help because of the sexual exploitation and rape that happened during the political protests ... The gender aspect has to be discussed because when internet shutdowns happen, women are the first victims of sexual exploitation, of hunger, and they are the first caregivers and are responsible for their families ... women need to contact diaspora communities to get money, medicine, and stuff like this. So because women are the first caregivers, they are the first to be affected by internet shutdowns [because they cannot use ICTs to communicate with service providers].[73]

The UN has made important strides toward taking internet shutdowns seriously, with the UN Special Rapporteur on Freedom of Peaceful Assembly and Association stating that internet shutdowns fail to meet Article 21 of the International Covenant on Civil and Political Rights because they undermine the right to peaceful assembly.[74] However, the UN lacks mechanisms that would require states to develop regulations that administer the compliance of digital tools with human rights.[75] Despite increasing research that documents the problematic effects of internet shutdowns and louder condemnations from civil society actors, policy responses remain limited. In certain cases, liberal democracies have responded with economic sanctions, such as in response to the 2019 Belarus presidential election and 2020 Myanmar parliamentary elections.[76] However, these sanctions were not an exclusive response to internet shutdowns but to the rampant human rights abuses that followed unfree and unfair elections. Whether internet shutdowns are a sufficient condition to trigger economic sanctions—or whether they should trigger sanctions—remains an open question that policymakers may want to consider, especially since sanctions can result in the further deterioration of human rights.

The international community has re-emphasized its commitment to tackling international cybersecurity. However, technology-facilitated violence that disproportionately impacts women—including doxing, the release of so-called "revenge porn," stalkerware, and harassment—are often outside the purview of international security discussions. A 2016 assessment of 20 national cyber security strategies found that all studied countries focused on securing cyberspace against malicious actors, with most countries, "especially Canada, USA, UK, Germany, Netherlands," perceiving the threat to "revolve around organized cybercrimes, state-sponsored attacks, cyber terrorism, unauthorized access to and interception of digital information, electronic forgery, vandalism and extortion."[77] Though nascent, there is now promising global coordination on issues that fall under the category of "cybercrime" even if these issues are not necessarily conceptualized as "international security." For example, Interpol has partnered with civil society and commercial partners to improve "the detection and mitigation of stalkerware" and raise "broader awareness of the issue alongside organizations that work directly with victims and survivors."[78] Given increased global coordination on gender-specific vulnerabilities, the call from civil society actors to make international (cyber)security "human-centered," and the introduction of gender norms into the OEWG, we may be witnessing a shift where cyber vulnerabilities are simultaneously understood to be social vulnerabilities. Internet shutdowns, however, occupy a tense position within this discussion because rather than being perpetuated by individuals, they are perpetuated by states. Consequently, internet shutdowns raise concerns long discussed in international relations about state sovereignty and foreign influence.

Policy recommendations have so far focused on creating programs that narrow the gender digital divide, cultivating legal remedies for marginalized groups that have experience ICT-enabled harassment and violence, protecting privacy-enhancing ICTs from policies that would degrade their effectiveness and prevent advocates from reaching their constituents, and increasing women's representation in technical and policy fields. I offer two recommendations designed to support research efforts in the ICT and international security field.

Recommendation 1: Policymakers should be encouraged to collect statistics that are disaggregated by gender. Depending on state capacity and resources, statistical data collection can be challenging and burdensome. Nevertheless, coordination and resource sharing between member states offer potential pathways for data collection that meet the necessary standards for the evidence-based policy formation. Currently, academics and NGOs are responsible for this task, resulting in a patchwork of important studies authored in different countries but that are nevertheless ill-suited for analyzing macro-level global trends that capture women's experiences with digital ICTs.[79]

Recommendation 2: As explained above, scholarship on gender and international security has noted the positive relationship between state stability and women's rights. However, little research currently exists that examines the relationship between the gender *digital* divide and international peace and stability. States should consider investing in research that examines whether women's access to digital ICTs is a positive indicator for international peace and stability. As states attempt to mainstream gender into their foreign policy activities, this type of research exploration can highlight where resources should be allocated to support international stability that centers human rights.

While the end of the Cold War and the international consensus on human security have caused states to rethink "what counts" as security, the absence of gender in digital ICTs and international security should be further reconciled. In particular, the link between the gender digital divide and global peace and stability provides a route for exploring whether women's improved access to ICTs can also have positive transnational effects on the international community. That being said, the most important argument for the inclusion of gender into the international security agenda is that women deserve to be full members of their political and social communities. The involvement of women in international security is not only good policy decision, but the right thing to do.

Notes

1 I had the opportunity to attend the OEWG meeting in December 2019 while working on an independent research contract with Global Affairs Canada. The culmination of that contract was a published report called *Making Gender Visible in Digital ICTs and International Security*. Certain parts of that report are expanded on here. My views do not represent Global Affairs Canada or any department in the Government of Canada.

2 Colliers and Lakoff, "The Vulnerability of Vital Systems: How 'Critical Infrastructure' Became a Security Problem," in *The Politics of Securing the Homeland: Critical Infrastructure, Risk and Securitisation*, edited by Myriam Dunn and Kristian Sobey Kristenson (New York: Routledge, 2008), 14.

3 Public Safety Canada, *National Strategy for Critical Infrastructure*, (Ottawa: Government of Canada, June 26, 2009).

4 Extensive debate surrounds "human security," with some researchers distinguishing between "broad" and "narrow" definitions. However, all versions of human security shift the security referent from the state to individual humans. Some definitions now also include the security of political communities. The 1994 definition popularized by the United Nations adopts a "broad" view of human security, which consists of freedom from fear, want, and indignity. For more discussion on the debate, see: Martin, Mary, and Taylor Owen (eds.), *The Routledge Handbook of Human Security*, (New York: Routledge, 2013).

5 Renata Hessmann Dalaqua, Hessman Renata, Kjølv Egeland, and Torbjørn Graff Hugo, *Still Behind the Curve*, (Geneva: UNIDIR, 2019), 10.

6 This report assumes that references to women include transgender women and girls and cisgender women and girls. However, vulnerabilities are often exacerbated due to age, gender identity, and expression. Notably, girls often experience different and more severe risks than women due to their childhood status.

7 In 2019, I was commissioned by Global Affairs Canada to write a paper on the gendered implications of digital ICTs in the context of international security. For that paper, which featured a case study on internet shutdowns, I conducted small-n interviews with five individuals for the purpose of understanding on-the-ground

conditions for women when internet access is cut, a topic that still remains underexplored. Given that the paper was designed for policymakers, the original project does not feature a section on the theoretical implications of my chosen methodology. A few, though not all, of the quotes that were used in the previous project have been shared again here when the discussion benefitted from interviewee insights.

8 Valerie Hudson, Mary Caprioli, Bonnie Ballif-Spanvill, Rose McDermott, and Chad F. Emmett, "The Heart of the Matter: the Security of Women and the Security of States," *International Security* 33, no. 3 (2009): 7–45.

9 Ian R. Gibson and Betty A. Reardon, "Human Security: Toward Gender Inclusion, in *Protecting Human Security in a Post 9/11 World,* edited by Shani G., Sato M., and Pasha M.K. (London: Palgrave Macmillan, 2007,) 52.

10 There is another possible explanation: civil society members know that some member states are averse to the language of human security and may have opted to use "human centered" for the sake of political expediency. However, this explanation would require further investigation and is currently untested.

11 Michael N Schmitt, ed., *Tallinn Manual 2.0 on the International Law Applicable to Cyber Operations,* (Cambridge: Cambridge University Press, 2017), 195.

12 Anne Marie Goetz and Rob Jenkins, "Agency and Accountability: Promoting Women's Participation in Peacebuilding," *Feminist Economics* 22, no. 1 (2015): 219.

13 Laurel Stone, "Women Transforming Conflict: A Quantitative Analysis of Women Peacekeeping," Preprint, May 13, 2014, SSRN: https://ssrn.com/abstract=2485242; Goetz and Jenkins 2015, 217.)

14 Hudson et al., "The Heart of the Matter," 41.

15 Government of Canada, "Gender Equality: A Foundation for Peace. Canada's National Action Plan," October 1, 2021, https://www.international.gc.ca/world-monde/assets/pdfs/cnap-eng.pdf; Mary Caprioli and Peter F. Trumbore (2003) note that states with higher levels of gender equality are more likely to "appeal to international organizations in their dispute resolution efforts, as both male and female leaders would have to appeal to broad constituencies that include feminine values" (2003, 199). Women voters have been found to prefer norms and values that Caprioli describes as having a "pacifying effect" on foreign policy (*ibid*, 197). These norm preferences are not necessarily biologically motivated. Women may be more likely to endorse these types of values due to social upbringing. See: Mary Caprioli and Peter F. Trumbore, "Identifying 'Rogue' States and Testing their Interstate Conflict Behavior," *European Journal of Political Science* 9, no. 3 (2003).

16 Caprioli and Trumbore, "Identifying 'Rogue' States," 10.

17 Giovanni De Gregorio and Nicole Stremlau, "Internet Shutdowns and the Limits of the Law," *International Journal of Communication* 14, no. 1 (2020): 4225.

18 Adrien Shabaz and Allie Funk, "The Freedom of the Net 2019: The Crisis of Social Media," *Freedom House,* (2019), 2.

19 Samuel Woodhams and Simon Migliano, "The Global Cost of Internet Shutdowns," *Top10VPN,* Jan 3, 2021, https://www.top10vpn.com/cost-of-internet-shutdowns/.

20 Surfshark, "Social Media Censorship Tracker," *Surfshark.com,* June 8, 2021, https://surfshark.com/social-media-blocking.

21 Surfshark, "Social Media Censorship Tracker."

22 Netblocks, "Internet disrupted in Myanmar amid apparent military uprising," *Netblocks.com,* January 31, 2021, https://netblocks.org/reports/internet-disrupted-in-myanmar-amid-apparent-military-uprising-JBZrmlB6.

23 Marianne Diaz Hernandes, Rafael Nunes, Felicia Anthonio, and Sage Cheng, 2021, "#KeepItOn Update: Who is shutting down the internet in 2021?" *Access Now,* June 7, 2021, https://www.accessnow.org/who-is-shutting-down-the-internet-in-2021/.

24 Lisa Parks and Rachel Thompson, "The Slow Shutdown: Information and Internet Regulation in Tanzania From 2010 to 2018 and Impacts on Online Content Creators," *International Journal of Communication* 14, no. 1 (2020): 4288.

25 Shayera Dark, "Strict New Internet Laws in Tanzania Are Driving Bloggers and Content Creators Offline," *The Verge*, July 6, 2018, https://www.theverge.com/2018/7/6/17536686/tanzania-internet-laws-censorship-uganda-social-media-tax.

26 Agence France-Presse, "Social Media Use Taxed in Uganda to Tackle 'Gossip.'" *The Guardian*, June 1, 2018, https://www.theguardian.com/world/2018/jun/01/social-media-use-taxed-in-uganda-to-tackle-gossip.

27 Kenya, Zambia, Mozambique, and Benin have also introduced social media taxes. See: Lisa Park and Rachel Thompson, "Slow Shutdowns," 4290.

28 Amy Cattle, "Digital Tahrir Square: An Analysis of Human Rights and the Internet Examined through the Lens of the Egyptian Arab Spring," *Duke Journal of Comparative and International Law* 26 no. 1 (2016): 434.

29 Berhan Taye and Sage Cheng, "The State of Internet Shutdowns," *Access Now*, July 8, 2019, https://www.accessnow.org/the-state-of-internet-shutdowns-in-2018/.

30 Hernandes et al., "#KeepItOn."

31 Deborah Brown and Allison Pytlak, "Why Gender Matters in International Cyber Security," *The Association for Progressive Communications,* April 21, 2020, 3, https://www.apc.org/en/pubs/why-gender-matters-international-cyber-security.

32 Sarah Shoker, *Making Gender Visible in Digital ICTs and International Security: A Report Commissioned by Global Affairs Canada,* (New York: UNODA, 2020): 18. https://front.un-arm.org/wp-content/uploads/2020/04/commissioned-research-on-gender-and-cyber-report-by-sarah-shoker.pdf

33 Amina Mohammed, "Better Laws, Data Essential for Tackling Cyberabuse, Growing Digital Gender Gap, Deputy Secretary-General Tells Event on Ending Online violence against Women," *United Nations Press Release,* March 14, 2018, https://www.un.org/press/en/2018/dsgsm1142.doc.htm.

34 Antonio and Tufley, 2014, 675.

35 See: Naeema Al Gaseer and Gwen Brumbaugh Keeney, "Status of Women and Infants in Complex Humanitarian Emergencies," *Journal of Midwifery and Women's Health*, (2004): 7–14; Jennifer Rumback and Kyle Knight, "Sexual and Gender Minorities in Humanitarian Emergencies," in *Issues of Gender and Sexual Orientation in Humanitarian Emergencies*, edited by Larry W. Roeder Jr. (Switzerland: Springer Nature, 2014): 33–74; Myriam Dunn Cavelty and Kristian Sobey Kristenson (eds.), *Security the Homeland: Critical Infrastructure, Risk, and (in)Security,* (New York: Routledge, 2008).

36 Interview 1, 2019.

37 J. Clement, "Global Internet Usage Rates 2019, by Gender and Region," *Statista*, January 28, 2020, https://www.statista.com/statistics/491387/gender-distribution-of-internet-users-region/.

38 World Wide Web Foundation, "Women's Rights Online: Closing the Digital Gender Gap for a More Equal World," *Web Foundation* (2020), 4, http://webfoundation.org/docs/2020/10/Womens-Rights-Online-Report-1.pdf.

39 World Wide Web Foundation, 2020, 10.

40 Mona Lena Krook and Juliana Restrepo Sanin, "Violence Against Women in Politics. A Defense of the Concept." *Politica y gobiern* 23, no. 1 (2016): 465.

41 J. Clement, "Global Internet Usage Rates."

42 Rajat Kathuria, Mansi Kedia, Gangesh Varma, Kaushambi Bagchi, and Richa Sekhani, "The Anatomy of an Internet Blackout: Measuring the Economic Impact of Internet Shutdowns in India," *Indian Council for Research on International Economic Relations*, 2018, https://think-asia.org/handle/11540/8248.

43 Henry Lancaster, *Myanmar (Burma)-Telecoms, Mobile, and Broadband-Statistics and Analyses*, (Sydney: BuddeComm, 2020), https://www.budde.com.au/Research/Myanmar-Burma-Telecoms-Mobile-and-Broadband-Statistics-and-Analyses.

44 Audrey Hingle, "What is the Digital Divide? Mozilla Explains," *Mozilla Foundation*, June 22, 2021, https://foundation.mozilla.org/en/blog/what-is-the-digital-divide-mozilla-explains/.

45 Peter Hunt, "Myanmar's Unsustainable Social Media Shutdown," *The Diplomat*, April 12, 2021, https://thediplomat.com/2021/04/myanmars-unsustainable-social-media-shutdown/.

46 For example, Psiphon, an internet censorship circumvention tool, has gained tremendous popularity in Belarus due to the internet shutdowns that have accompanied the 2020 presidential election. See: Adrian Hamacher, "Tor and Psiphon Activity Surges in Protest-Stricken Belarus," *Decrypt*, Aug 12, 2020, https://decrypt.co/38443/tor-and-psiphon-activity-surges-in-protest-stricken-belarus.

47 Oxfam International, "Why the Majority of the World's Poor Are Women," Oxfam International, June 10, 2021, https://www.oxfam.org/en/why-majority-worlds-poor-are-women.

48 Hingle, "What is the Digital Divide?" 2021.

49 Digital redlining is the perpetuation of discrimination between groups using digital tools, often through delivering or withholding digital technologies on the basis of geographic location.

50 Hussain Arif Nadaf, "'Lockdown within a Lockdown': The 'Digital Redlining' and Paralyzed Online Teaching during Covid-19 in Kashmir, a Conflict Territory," *Communication Cultural Criticism*, April 8, 2021, https://www.ncbi.nlm.nih.gov/pmc/articles/PMC8083513/.

51 Asmita Bakshi, "India Is the Internet Shutdown Capital of the World," *Livemint*, December 8, 2019, https://www.livemint.com/mint-lounge/features/inside-the-internet-shutdown-capital-of-tHe-world-11575644823381.html.

52 UN Women, "The Shadow Pandemic: Violence against Women during Covid-19," UN Women, June 18, 2021, https://www.unwomen.org/en/news/in-focus/in-focus-gender-equality-in-covid-19-response/violence-against-women-during-covid-19.

53 Puja Changoiwala, "The Doctors Navigating Covid-19 with no Internet," *The BMJ*, April 7, 2020, https://covid-19.conacyt.mx/jspui/bitstream/1000/2853/1/1102825.pdf.

54 Changoiwala, "The Doctors," 2020.

55 Julie Ballington, Gabrielle Bardall, and Gabriella Borovsky, *Preventing Violence against Women in Elections: A Programming Guide* (New York: UNDP and UN Women, 2017).

56 Interview 2, 2019.

57 Interview 4, 2019.

58 Interview 4, 2019.

59 Giovanni De Gregorio and Nicole Stremlau, "Internet Shutdowns and the Limits of the Law," *International Journal of Communication* 14, no. 1 (2020): 4230.

60 De Gregorio and Stremlau, "Internet Shutdowns," 4229.

61 Flynn, Meagan, "No One Who Watched New Zealand Shooter's Video Live Reported it to Facebook, Company Says," *Washington Post,* March 19, 2019, https://www.washingtonpost.com/nation/2019/03/19/new-zealand-mosque-shooters-facebook-live-stream-was-viewed-thousands-times-before-being-removed/.

62 Public Safety and Emergency Preparedness Canada, "Statement from Minister Goodale following the G7 Interior Ministers Meeting," *Newswire.ca,* April 5, 2019, https://www.newswire.ca/news-releases/statement-from-minister-goodale-following-the-g7-interior-ministers-meeting-842283557.html.

63 Interview 3, 2019.

64 Tom Embury-Dennis, "Extinction Rebellion: London Tube Wifi Shut Down by Policy in Attempt to Disrupt Climate Change Protestors," *Independent*, April 17, 2019, https://www.independent.co.uk/news/uk/home-news/london-tube-wifi-down-internet-not-working-underground-protest-extinction-rebellion-a8873681.html.; Eleanor Marchant and Nicole Stremlau, "The Changing Landscape of Internet Shutdowns in Africa—Introduction," *International Journal of Communication* 14, no. 1 (2020): 4218.

65 David Kravets, "San Francisco Subway Shuts Down Cell Phone Service to Foil Service; Legal Debate Ignites," *Wired*, August 15, 2011, https://www.wired.com/2011/08/subway-internet-shuttering/.

66 Wolchover, Natalie, 2011. "How Did BART Kill Cellphone Service?" *Scientific American*, August 16, 2001, https://www.scientificamerican.com/article/how-did-bart-kill-cellpho/.; De Gregoria and Stremlau, "Internet Shutdowns," 2020, 4225.

67 Kravets, "San Francisco Subway," 2011, and James Vincent, "UK Police Shut Down Wi-fi in London Tube Stations to Deter Climate Protestors," *The Verge,* April 17, 2019, https://www.theverge.com/2019/4/17/18411820/london-underground-tube-wi-fi-down-shut-off-protests-extinction-rebellion.

68 Deniz Duru Aydin, "Five Excuses Governments (ab)use to Justify Internet Shutdowns," Access Now, October 6, 2016, https://www.accessnow.org/five-excuses-governments-abuse-justify-internet-shutdowns/.

69 Interview 3, 2019.

70 Interview 1, 2019.

71 Djaouida Siaci, "The Mass Rape of Rohingya Muslim Women: An All Out War against All Women," *Middle East Institute,* September 29, 2019, https://www.mei.edu/publications/mass-rape-rohingya-muslim-women-all-out-war-against-all-women; Alex Warofka, "An Independent Assessment of the Human Rights Impact of Facebook in Myanmar," Facebook.com, November 5, 2018, https://about.fb.com/news/2018/11/myanmar-hria/.

72 For a more extensive discussion on the ways that women's rights organizations have responded to digital insecurity, see: Shoker, "Making Gender Visible in Digital ICTS," 2020.

73 Interview 1, 2019.

74 United Nations Human Rights Council, *Ending Internet Shutdowns: A Path Forward*, (Geneva: General Assembly, Forty-Seventh Session, 2021), 4, https://undocs.org/A/HRC/47/24/Add.2.

75 Marie Lamensch, "The Cost of an Internet Shutdown," Centre of International Governance Innovation, March 9, 2021, https://www.cigionline.org/articles/cost-internet-shutdown/.

76 The Associated Press, "Canada, U.K. Impose Sanctions on President of Belarus and Seven Others," *CBC.ca,* September 29, 2020, https://www.cbc.ca/news/world/canada-u-k-imposes-sanctions-president-belarus-7-others-1.5742981; Bloomberg News, "How Myanmar's Coup Put Democracy on the Back Burner again," *Bloomberg* February 1, 2020, https://www.bloomberg.com/news/articles/2021-02-01/why-myanmar-s-stop-start-democracy-is-back-on-hold-quicktake.

77 Narmeen Shafqat and Ashraf Masood, "Comparative Analysis of Various National Cyber Security Strategies," *International Journal of Computer Science and Information Security* 14, no. 1 (2016): 132.

78 Interpol, "Taking a Stand against Online Stalking," Interpol, April 21, 2021, https://www.interpol.int/en/News-and-Events/News/2021/Taking-a-stand-against-online-stalking.

79 For more discussion on sex-disaggregated gender statistics and the gender digital divide, see: Sarah Shoker, *Making Gender Visible in Digital ICTs and International Security: A Report Commissioned by Global Affairs Canada*, (New York: UNODA, 2020).

PART II

4

ASEAN AND GENDERED VIOLENCE IN CYBERSPACE

Fitriani Bintang Timur

Introduction

In 2020 Southeast Asia, nearly 70% of the population has access to the internet,[1] including a 40 million increase in the number of users since the spread of COVID-19.[2] Despite the encouraging statistics on internet connectivity, such progress cannot be correlated with a proportionate level of digital awareness regarding data and privacy in cyberspace. Besides, when women have access to the internet, they are disproportionately targeted, enduring more online violence than men through a continuum of multiple, recurring and interrelated forms of gender-based violence (GBV).[3] The United Nations Population Fund (UNFPA) reported that for the period of October 2019–2020, the percentage of misogynistic tweets has increased 22,384% in Thailand, 935% in the Philippines, 140% in Singapore and 21% in Indonesia.[4] Concurrently, the percentage increase in tweets providing support for victims of online violence, while reaching a high of 112% in Thailand and 56% in Indonesia, observed a decline of −4% in the Philippines and −40% in Singapore.[5] These numbers empirically demonstrate both the increases in misogynistic threats and in GBV awareness, two of which are factors that should push Southeast Asian governments to develop more gender-inclusive standardisation.

Unified regional efforts to address ever-changing threats in cyberspace have been discussed, planned and carried out multilaterally through the Association of Southeast Asian Nations (ASEAN). The Southeast Asia region, much like other regions in the world, believes that digital transformation is the key to maximise value creation from technology's rapid growth and bring economic advancement for society.[6] However, this chapter argues that the region should also be wary of the potential gendered harms that cyberspace might bring and strive to foster regional cooperation despite the existing constraints of ASEAN

DOI: 10.4324/9781003261605-6

member states' (AMS) diversity of digital infrastructure, cyber capability and understanding of gender. It will focus on ASEAN for three reasons: First, the borderless nature of cybersecurity means that the perpetrator may sit in a certain jurisdiction, yet their action can affect many parts of the region, and therefore studies on a national scale would not be sufficient. Second, the AMS share similar threats, internet demography and trends that call for a unified approach to combat the rising threat. Third, ASEAN has a regional architecture with shared norms, joint commitments and a pool of resources for addressing GBV that can be further developed. These include the Declaration on the Elimination of Violence Against Women in the ASEAN Region (2004), ASEAN Regional Plan of Action on the Elimination of Violence Against Women (2015) and Joint Statement on Promoting Women, Peace and Security in ASEAN (2017).

The chapter first provides a brief background on online GBV in ASEAN countries to demarcate the sense of urgency for a unified regional approach. Subsequently, the writing applies the 2021 United Nations Institute for Disarmament Research's (UNIDIR) Gender Approaches to Cybersecurity that focus on the three pillars of Design, Defence and Response. This "Three Pillars Framework" is chosen because it provides a structured way to analyse GBV as a phenomenon. This Framework is considered relevant for Southeast Asia as the approach of analysing Design, Defence and Response in regional cybersecurity has been done before within literature[7] but they lacked gender perspective. Hence, it is hoped that this approach will provide a more inclusive analysis. The Framework is used to analyse AMS' response towards the increasing frequency of online GBV to argue that there is a need for a more gender-sensitive approach, and initiatives need to be taken to counter this rising threat. Literature reviews on ASEAN's approach towards online GBV and the increased number of online GBV during the COVID-19 pandemic will offer experts' opinions on the importance of data collection and understanding of online GBV's pervasiveness in creating society's awareness. As this chapter covers the region of Southeast Asia, it mainly discusses gendered violence in cyberspace directed towards women despite the reality that such violence not only affects them, but also affects diverse gender identities. The reason is that there are countries within ASEAN that have yet to openly endorse and protect the rights of people with diverse sexual orientation and gender identity and expression (SOGIE), which limits publication on GBV experienced by gender minorities and non-heteronormative peoples, including lesbian, gay, bisexual, transgender, intersex, queer/questioning and asexual (LGBTQ+) communities. Therefore, the caveat of this chapter is that it provides limited coverage of the gendered communities it discusses as it speaks mostly on women. To close, despite the challenges of member states' diversity on cyber capability and understanding of gender, this chapter recommends that ASEAN formulates a regional plan of action to address online GBV.

Status quo of online GBV

This chapter applies the concept proposed by Millar, *et al.*, which defines cybersecurity as the "prevention and mitigation of malicious interference with digital devices and networks".[8] Cybersecurity refers to the overall environment or global domain in which internet usage is facilitated. Ensuring safety in cyberspace is cybersecurity.[9] Meanwhile, this chapter defines GBV as "an act committed, abetted, or aggravated, in part or fully, by the use of ICTs, such as mobile phones, the internet, social media platforms, and email".[10] One of the elements of GBV is gender-based cyberviolence that targets at a person because of their gender, and this may take the form of threats with physical or sexual harm, sexist content, stalking, bullying, revealing personal information or private images without consent, or posting abusive comments which negatively impact women's experience of digital access.[11] Facing the growing risk of online GBV in cyberspace requires a specific approach to ensure "gender-responsive" and "gender-sensitive" policies.

Women, minorities and LGBTQ+ communities are commonly subjected to, and experience discrimination and injustice offline, and increasingly find themselves in similar situations online; demonstrating that disparities, powered by stereotypes and social norms, continue to exist. Particularly in Southeast Asia, the lack of sensitivity and regulation on GBV acts on social media platforms such as Twitter and Facebook, which combined with anonymity, has created a culture of impunity,[12] allowing perpetrators to feel less responsible for their misconducts online. In 2020, within the Asia Pacific, the number of misogynistic narratives on Twitter reached 49%, with 27% labelled misogynistic profanity and 24% as misogynistic humour.[13] It is these structures, along with the rapid digital adoption brought about by the COVID-19 pandemic's constraints to face-to-face meetings, that have allowed cyber GBV to increase by 600% during the pandemic; with the most common forms including sexist and misogynistic comments, gender-based hate speech, identity theft, unsolicited nude images or online sexual solicitations, sexually explicit content and sextortion.[14]

The AMS experienced various challenges to overcoming cybercrimes. Numerous fraudulent websites in Thailand mimic government institutions to acquire personal information, with women and minority groups, including LGBTQ+ communities, placed in more vulnerable position as their leaked address enable stalkers and criminals to target them.[15] Myanmar experienced an increase in internet-assisted human trafficking used for scouting and grooming victims, forcing more girls and women into involuntary labour, sexual servitude and forced marriage.[16] In the Philippines, nearly one in two children aged 13 to 17 had experienced online GBV[17] with a total of 280,000 cases of online sexual abuse against children.[18] Meanwhile, in Malaysia, research has shown that individuals affiliated with the LGBTQ+ community (whether belonging to or in support of) are more vulnerable to rape and sexual exploitation facilitated through dating apps.[19] In Indonesia, the National Commission on Violence

Against Women recorded 695 cases of online sexual harassment and abuse from January to October 2020 – double the amount of the year before.[20] These alarming increases in the level of online GBV demonstrate the amount of effort needed by the AMS to combat the challenges and work towards prevention to ensure inclusive safety and security for all, including in cyberspace.

Even though rapid digitalisation offers many positive economic and social benefits, it has also enabled online GBV. The digital realm's provision of anonymity, ease of access towards other person's personal data and the vast spread of information has put women and other vulnerable groups in an even more assailable position. Anonymity can be a double-edged sword, where it helps vulnerable groups or communities express dissenting opinion, but anonymity also enables perpetrators to perpetuate discrimination without fearing backlash. Until today, women still face difficulties guarding themselves against, and responding to online GBV, especially when laws in many countries fail to criminalise such acts. This is coupled with law enforcement and criminal justice systems that are not yet equipped to trace, arrest and punish perpetrators. Such conditions remain similar across the Southeast Asian region, where 2020 research findings by UN Women showed that Malaysia and the Philippines scored 4 and 5 out of 5 respectively, with 5 being the worse, for occurrences of distribution of non-consensual sexually explicit content, sextortion, doxing, online harassment, discrimination and online threat of violence both sexual and non-sexual.[21] On the policymaking front, the lack of gender representation in the cyber or digital industry poses a challenge to the provision of adequate cyber security for women, girls and vulnerable groups like the LGBTQ+ community. Additionally, there is an increased risk of online GBV towards women, girls and other vulnerable groups. This stems from variances in awareness both amongst social media companies that track and prevent online GBV using local Southeast Asian languages, and differing levels of awareness among users in the region.[22] One of the commonly known cyber security concerns in the region is data breaches, nevertheless gender dimension is rarely used in examining the impact of such incident. Examples of data breaches in Southeast Asia in the recent years include the 2018 Singhealth data leaks in Singapore; the 2019 Thailand and Vietnam Toyota data breaches; as well as the Indonesia's 2020 Tokopedia leaks and State Health Data sold on the black market, all showing the vulnerabilities of data protection by ASEAN governments and businesses' alike. It is important to take note that data utilisation and exploitation never take place in gender-neutral settings. Even though data leaks do not target specific genders, they impact women and gender minorities more severely due to existing structural inequality and discrimination. Despite the lack of gender analysis and sex-disaggregated data when it comes to data breaches, anecdotal evidence of the doxing of female journalists, such as that experienced by Ika Ningtyas from Indonesia and Maria Ressa from the Philippines, is aplenty.[23] As the case study of Maria Ressa and other women journalists who experienced online violence as a result of their public professional work show, data breaches and intentional disclosure of information enable the targeting of specific gender.[24]

Literature discussing online GBV that covers the scope of Southeast Asia remain limited, with only two (Sey, 2020 and Swe, 2019) portraying analysis on initial conditions of ASEAN's online GBV, and three others (Jatmiko, *et al.*, 2020; Dlamini, 2021; UN Women, 2020(a)) discussing the occurrences of online GBV during the COVID-19 pandemic. Araba Sey's work focuses on gender digital equality across ASEAN, particularly in the field of digital economy, and Sey includes an analysis pointing out the dark side of digital economy: gender-based cyberviolence. Sey points out ASEAN's recognition of online GBV through their Declaration of Elimination of Violence against Women and Elimination of Violence against Children in ASEAN and the ASEAN Regional Plan of Action on the Elimination of Violence against Children. However, the region has not responded substantively to the problem and online harassment has not been well-documented. Her findings, though mainly revolving around the high percentage of cyberbullying towards girls, concluded that cyberviolence might hinder vulnerable groups from experiencing and/or maximising benefits of digital technologies.[25]

On the other hand, Swe (2019) focused his analysis on the need for regional approach to cyberbullying as one of the most common online GBV within ASEAN. Nationally, countries have various regulations due to differences in social context and rate of cyberbullying; but Swe argued that ASEAN needed to scale up its efforts to combat cyberbullying. He highlighted the role of the ASEAN Commission on the Rights of Women and Children (ACWC) in providing technical assistance to the AMS to deal with cyberbullying through workshops for teachers and reinforcing or motivating new initiatives of an anti-cyberbullying programme.[26] Both Sey and Swe assessed regional efforts and established grounds on the general conditions of cyber violence and/or online GBV in ASEAN. Sey argued that regional efforts have been limited, and Swe provided recommendations on how regional efforts can be intensified.

Specific to the COVID-19 context, a study by Jatmiko *et al.* (2020) measured harassment on social media that created domino effect on victims/survivors' mental and physical health amidst increased internet dependence during the pandemic.[27] The study also highlights the problematic approach that social media platforms have when it comes to GBV issues. Social media platforms under-publicise the prevalence of GBV and the methods that victims/survivors can pursue when experiencing online GBV. This causes perpetrators to be unaware of the consequences of online GBV and indirectly drive victims to choose not to report due to unclear methods to address the issues they face. Consequently, victims/survivors remain reluctant to report the online GBV and sexual violence they experienced, causing statistics of complaints and reports to be deemed minimal and insignificant.[28]

Similar findings are also confirmed by Dlamini (2020) who observed the increase in occurrence of GBV towards women and girls due to the pandemic in the physical space and also in the digitised world.[29] Increase of online GBV inhibits women from accessing the internet, especially those who are identified as part of the highly targeted groups: young women; women from ethnic or racial minorities; indigenous women; lesbian, bisexual and transgender persons; women with disabilities; women human rights defenders; journalists; women from marginalised groups; and those facing multiple and intersecting forms of discrimination. Dlamini recommended, that on a national level, the government should address the issue of gender inequality through the implementation of gender budgeting, as well as for civil society organisations to raise the awareness to take actions. Dlamini's recommendations are built on earlier literatures on online GBV, including that of Suzor, *et al.* (2019), which argues the necessity for technology companies to prevent and combat abuse across their networks, and Lewis, *et al.* (2017) who contend that GBV is a form of abuse and therefore needs to be addressed seriously.[30]

Meanwhile, UN Women 2020 report for Asia and the Pacific region noted a sharp rise in the number of online sexual exploitation and abuse towards women and girls, and argued that such acts are accompanied by increased access to child sexual abuse material displayed on illegal websites.[31] There is a concern that online GBV can be accompanied by or lead to in-person sexual and GBV against women and girls. [32] The report also mentioned the use of the internet as a recruitment platform for trafficking and smuggling migrants, with Myanmar being a pertinent example. On the other hand, the governments in the region tightened censorship to battle against disinformation which curtailed women activists' and journalists' freedom of expression online, as observed in Cambodia, Malaysia, Myanmar, the Philippines, Thailand and Vietnam.[33] The 2020 report by UN Women proposed to strengthen cybersecurity through the Women, Peace and Security agenda, and address the gender gap within the workforce. The latter recommendation is used in this chapter's analysis.

The literature review demonstrated that the pandemic rapidly increased GBV perpetrated through online platforms. Rising cases of in-person GBV due to lockdowns should not overshadow the urgent need to address online GBV as the latter continues to rise during the pandemic. So far, existing studies remain focused on microscale data gathering through interviews with survivors (Jatmiko, *et al.*, 2020) or general survey to provide a general understanding of the different forms of online GBV during the pandemic (Dlamini, 2020; and UN Women, 2020(a)).

This chapter seeks to address the absence of literature applying the Three-Pillar Framework – Design, Defence and Response – to improve gender sensitivity and responsiveness, with a focus on the ASEAN region. The area scope is applied to focus on the group of states with shared norms, mostly in relatively similar conditions of being developing countries aspiring to increase their welfare through the application of technology and have existing joint commitments in addressing GBV.

The first pillar (Design) refers to the building of security capabilities into a socio-technological system to reduce the surface area for attack and prevent whole classes of vulnerability from being formed. This pillar seeks to prevent and mitigate malicious interference by modelling threats and designing models to counter these threats using a more gender-sensitive and responsive standards. It focuses on the existing terms and conditions, and standards relating to gender-sensitivity and gender-responsiveness within the countries of ASEAN. The second pillar (Defence) concerns the reduction of risk, the identification of vulnerabilities and the amelioration of potential harms through constant improvements of the cybersecurity framework. The Defence Pillar assesses the extent to which the governments and/or civil organisations are committed to protect and encourage the participation of women in the cyberspace and cyber-security industry. The third pillar (Response) refers to how states respond to incidents of intrusions or disruptions of digital devices and networks. It heavily relies on the support of legal measures and legislation to deal with cybercrime; however, there are still gaps in proper legal responses, most notably where women and LGBTQ+ communities are concerned.[34]

Framework and analysis

As previously mentioned, scholars such as Khanisa (2013) and Heinl (2014) conducted analyses on regional cybersecurity using the three pillars of Design, Defence and Response. However, none of them utilised a gender-based perspective. Therefore, it is important to revisit the issue using a gender-sensitive lens.

First pillar: Design

Currently, cybersecurity and cyberspace standards are governed by a wide range of international bodies and a mostly male/masculine dominated perspective, with very limited participation by regional bodies and an almost absent gender perspective. In an effort to include gender-sensitivity and gender-responsiveness as part of standardisations, the International Telecommunication Union (ITU) formed its Women in Standardization Expert Group (WISE), and the International Organization for Standardization (ISO) developed a Gender Action Plan. The United Nations Economic Commission for Europe (UNECE) published a declaration on "Gender Responsive Standards and Standards Development"[35] that encapsulates combined effort towards the development of gender inclusive standardisation, signed by 72 countries and standard-development organisations.[36] Reflecting upon these, it is not new that approaches on cybersecurity regulations, frameworks and norm-setting are discussed by regional and international organisations. However, although international bodies hold the role as a universal and global platform for discussions and dialogues on cybersecurity, they find it rather difficult to cater to the specific concerns and needs of every single nation. This is where regional organisations like ASEAN are expected to fulfil their role. Having relatively similar internet growth trends, demographics, threats and even opportunities from the rapid cyber development, the AMS would benefit from shared ASEAN cybersecurity and cyberspace standards. With similar goals and threats, ASEAN enables member states to cooperate and coordinate best practices for regional action.

ASEAN already has shared commitments for combating cybercrime,[37] but they do not include aspects of gender-responsiveness and gender-sensitivity for responding to the rising concerns of online GBV. Even before the COVID-19 pandemic, national governments of the AMS struggled to develop a basic set of standards on cybersecurity, with a clear absence of gender-sensitivity and gender-responsiveness.[38] Not having a developed framework for cybersecurity management and protection against cybercrime is one of the crucial challenges the region needs to counter. Majority of the ASEAN countries' cybersecurity design still heavily focus on military and corporate security measures instead of a human-centric cybersecurity approach, which sets human safety as the main subject of cybersecurity regulations. This calls for a regionally unified approach towards cybersecurity design standardisation which allows countries to coordinate based upon their shared threats and opportunities be it through regional norm-building or a regional action plan requiring member states to meet regional commitments on cybersecurity standardisation. At the time of writing, the region has developed the 2020 ICT Masterplan, the 2015 ASEAN Regional Forum Workplan on Security and the Use of ICT, ASEAN Framework of Digital Data Governance and ASEAN Framework on Personal Data Protection. However, there is a visible lack of incorporating gender-sensitive and gender-responsive approach to the design of cybersecurity in the region. Understanding this missed opportunity, it is recommended that ASEAN formulate a regional action plan to address GBV, learning from other regions such as the European Union (EU) with the EU Gender Action Plan III (2020) that includes the goal of protecting human rights, both online and offline, and ensuring a safe and secure cyber space, with standardised data protection such as EU General Data Protection Regulation (GDPR).

Second pillar: Defence

The following section discusses legal regulations and how the design of workforce participation can be improved to provide protection for women against GBV. While the everchanging nature of technology makes it impossible to develop a perfect cybersecurity design, countries should continue improving on their defence mechanisms guided by the Defence Pillar – that is to protect and encourage the participation of women in the cyber-industry. Analysis of the Defence Pillar includes (1) how regulations and legislations of the AMS' national governments protect and prevent against online GBV and (2) the participation of women in the cyber-industry allowing for a gender perspective in policymaking and decision-making. These measures will be analysed using a national-scale perspective to see the gap amongst countries, then determining how ASEAN would effectively bridge the gap.

Following the two prioritises of the Defence Pillar, protection and participation, besides ongoing efforts to make its tech sector more inclusive to women, the AMS can also focus on setting up regulation to provide better protection. Protection here refers to countries' national regulations on how they choose to prevent and regulate online GBV. Every Southeast Asian country possesses differing national laws governing their cyberspace. In Cambodia, the dissemination of images and conversations without consent, breach of professional secrecy, violation of secrecy and fraud are regulated under the Penal Code with applicable fines and imprisonment sentences. Laos possesses a Law on Combating and Preventing Cybercrime (no. 61/NA), primarily targeting harmful online actors seeking to do harm. Both Laos and Cambodia prohibit the disclosure of private confidential information. Myanmar's 2013 Telecommunication Law prevents the spread of private information gathered through online transactions.[39] Thailand's Cybersecurity Act and Personal Data Protection Act, and Vietnam's Law on Cyber-Information Security, focus on consumer data protections.[40] The Philippines' 2012 Cybercrime Prevention Act punishes cybercrime. Unfortunately, Indonesia's Electronic Information Law and Malaysia's Communication and Multimedia Act are not wholly restrictive.[41] In terms of citizen cyber-protection, the AMS vary greatly. The disparity calls for a unified approach with the help of their regional organisation, ASEAN, to coordinate, share best practices and raise the awareness of cybersecurity.

Another important factor of the defence pillar is the participation of women. The participation of women is vital because it allows for the inclusion of gender-based perspectives through the diversification of the workplace. Unfortunately, global statistics show that women's participation rates compared to men's remain abysmal, making up 15–20% of the total workforce in cybersecurity and even lower in the information security sector in 2020.[42] Within the United Nations Group of Governmental Experts, women's participation is, at average, 20%, which is still below the recommended 30% threshold for adequate influence.[43] To address these issues, several countries in the region have advocated for greater women's representation, for example, Indonesia with Indonesia's Women in Cybersecurity (IWCS) and Singapore's SG Cyber Women Program. These initiatives support greater women's representation in Southeast Asia's cyber industry as shown in Figure 4.1.

Within Southeast Asia, the disparities of women in technology remain a challenge (Figure 4.2). In the region, Thailand has been the most successful in educating women in technology (48%) and channelling them to work in the same field (42%). The Philippines and Malaysia have close percentages of women graduates in technology (48% and 46%, respectively), but it translates to only 35% women entering tech work.[44] Indonesia faces similar problems with a higher percentage of women receiving education in technology (35%) than women entering the workforce (22%).[45] The statistics raised a concern that a

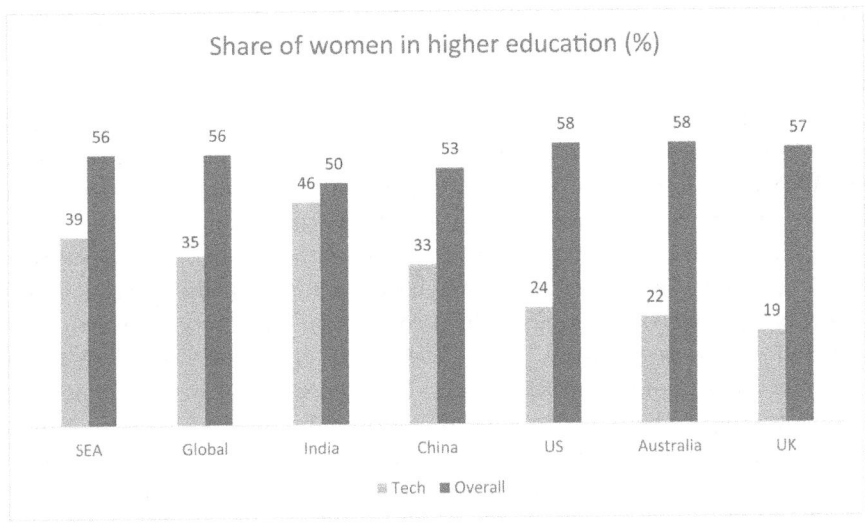

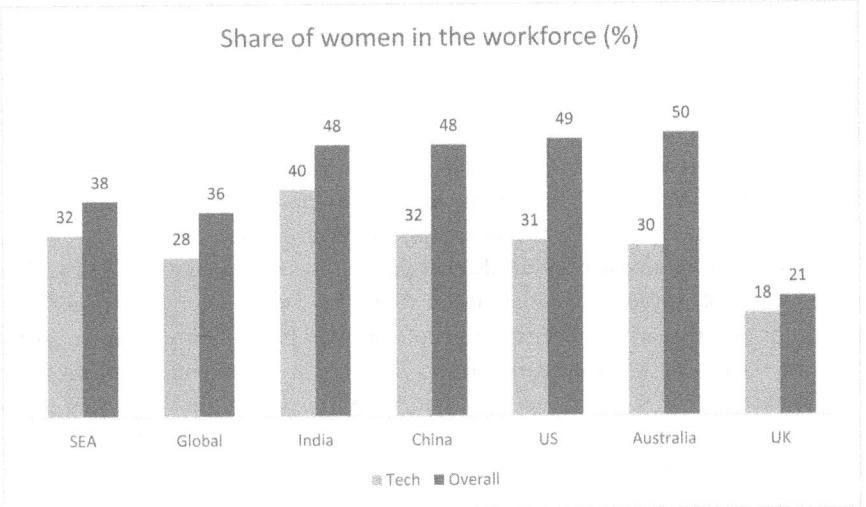

FIGURE 4.1 Women's Representation in Southeast Asia Compared with Other Areas

Source: Vaishali Rastogi, et al. (2020)

higher number of women receiving education in technology does not auto-matically translate to higher participation in relevant workforce, and therefore further study on the causes should be a pressing issue.

Interestingly, the 2020 data shows Singapore and Vietnam saw a relatively low share of women in technology-related majors (29% and 26%, respectively) but have higher shares of women working in technology (41% and 34%, respec-tively).[46] The likely reason for more women absorbed to careers in technology in the two countries is less due to industry inclusivity initiatives, but more because

FIGURE 4.2 Women's Representation in Tech Still Lags That of Other Industries in Southeast Asia

Source: Vaishali Rastogi, et al. (2020)

both Singapore and Vietnam have experienced booming tech sectors since the early 2010s that attracted women from non-tech backgrounds. [47] The numbers signify that while the number of women studying technology might increase, it does not mean that participation of women in the cyber-industry will increase, and vice versa. Though reasons may vary, data on the representation of women in technology should alert countries like the Philippines, which has the highest percentage of female graduates from technology education but low female representation in the workforce of the need to act. The Philippines needs to evaluate their cyber industry's recruitment process, trends in workforce and further analyse the anomaly, in order to address the issue by encouraging more women's representation in the industry.

Seeing the statistics of women obtaining education and working in technology, Singapore and Vietnam are quite successful in improving the skills of women to maintain relevance in the workforce and channelling them to the growing sector of technology. One notable example of Singapore's best practice is its Workforce Development Agency that uses a two-pronged approach of continuous education and job matching.[48] These differences in the region call for a knowledge- and strategy-sharing initiative if the region seeks to increase the participation of women in tech in the region to compete globally. However, instead of building regional reputation, countries in Southeast Asia often compete with one another in order to be seen as more advanced, and to get international technology companies set up their business and manufacture.[49]

Laws and regulations are among the most common and important measures a government can take to accomplish protection in cyberspace; yet, as previously mentioned, regional progress remains slow compared to the growth of cyber threats. There is a global push for policymakers to guide the direction of accelerating digital transformations so it can ensure equal opportunity and safety for all,[50] the situation is similar in Southeast Asia.[51] Women's participation is equally important, but, just like the development of law and regulation, progress varies.

Gender-responsive and gender-sensitive defence capabilities do not merely lie on women's representation in the cyber-industry, but also in the policymaking and law sector. Southeast Asia still lacks women's representation in policymaking. As noted by Inter-Parliamentary Union statistics, the proportion of women in parliament is still underrepresented. Per January 2021, Singapore has the highest percentage of women in the parliament (29%) and Indonesia has in the ministerial (17%), whilst Brunei has the lowest percentage of women in the parliament (9%) and the three countries of ASEAN – Thailand, Brunei and Vietnam have no women in their ministerial.[52]

Women's representation in non-cyber industry sector, especially in politics (parliament), enables a gender-responsive social structure which would open doors for women's representation in other fields. Especially in the cyber industry where men are still dominant, women's representation in politics would enable a structural change leading to representation in more specific fields through policymaking. ASEAN Sustainable Development Goals Baseline Report 2020 reported that the share of women in national parliaments was 19.6% in 2018 (Figure 4.3).[53] Overall, there has been positive progress in women's representation in policymaking from 2016 to 2018, creating a hope for more gender-responsive regulations. However, several countries possess higher percentages of women's representation compared to others in the region; this calls for a wider connected effort to achieve adequate cybersecurity defence capabilities throughout the whole region. Arguably, as women are best placed to identify their unique cybersecurity needs, including by taking the consideration of their lived experiences to the knowledge-base to shape cybersecurity, it has become a challenge that their representation remains low.[54] For example, the gap between Brunei Darussalam and Vietnam demonstrates the disparity and provides room for improvement, which specifically calls for ASEAN's role in advocating for regional response and unified action in pursuing better and equal cybersecurity defence system.

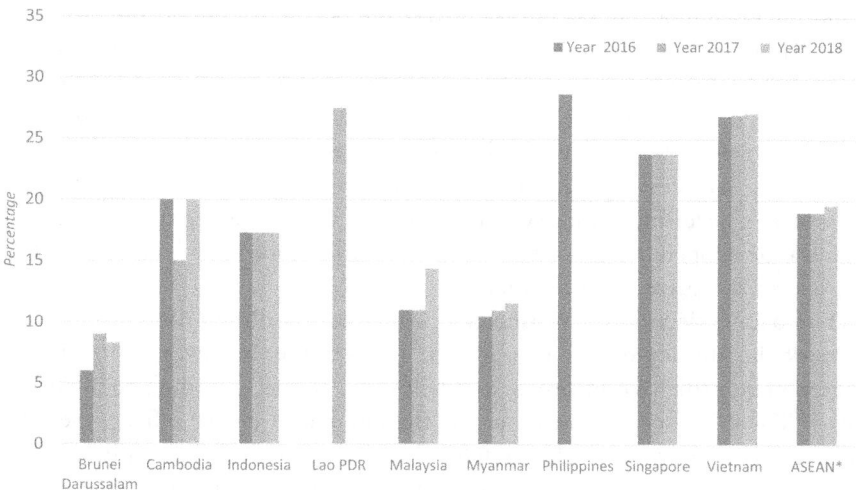

FIGURE 4.3   Proportion of Seats Held by Women in National Parliaments in ASEAN, 2016–2018

Source: ASEAN Sustainable Development Goals Indicators Baseline Report (2020)

Third pillar: Response

The last pillar observes the responses and measures taken to handle network intrusions, attacks, data breaches and other malicious cyber acts that are categorised into a hierarchy of priorities. The AMS benefit from online transactions, and financial technology is one of the sectors the AMS is focusing on. It is reasonable that national legal measures amongst the AMS prioritise addressing cybercrimes relating to economic and corporate transactions. However, increased usage of the internet does not only affect the economic sphere, but also social life, and cybercrime is not exclusive to the economic sector. The growing scope of cybercrime should alert the AMS to expand and have proper responses towards all kinds of cybercrime. Particularly, online GBV which is often left untouched, even though it is statistically proven to be a constant threat. The urgency of responding to online GBV lies in its long-term impact towards the fight against gender inequality. As a part of the Sustainable Development Goals (SDG) Goal 5 to "Achieve gender equality and empower all women and girls",[55] the attainment of gender equality requires legislative measures to prevent and respond to all kinds of gender discrimination. With the increased levels of internet activity due to COVID-19 along with increased online GBV, the AMS should have a proper response towards online GBV.

ASEAN's existing regional policy response to GBV includes the Declaration on the Commitments for Children in ASEAN that aims to protect children from all forms of violence and exploitation at the home, school and community; the ASEAN Regional Plan of Action on the Elimination of Violence Against Children 2016–2025 (RPAEVAC); the ASEAN Regional Plan for

Violence Prevention and Protection and the ASEAN Commission for the Promotion and Protection of the Rights of Women and Children (ACWC).[56] In terms of protection and support for victims and survivors, the ASEAN Regional Plan of Action on the Elimination of Violence Against Women includes services such as, but not limited to, medical and psychological care, counselling services, legal aid, interpretation and services, hotlines as well as assistance to accessing justice systems.[57] However, considering that this action plan was developed in 2016, updates and adaptations need to be made to better respond to the current developments.

The data collection on victims and instances of online GBV is necessary to track the efficacy of the Three-Pillar Programme, however, the ASEAN governments still have not recognised this as a priority, leading to a gap in data.[58] There is also a need to increase the number of justice officers specialising in cyberviolence so that gendered cybercrime cases and online GBV can be adequately addressed. Globally, the numbers of such justice officers are increasingly rare,[59] whilst regionally, there has yet to be any visible discussion at the ASEAN level, or at the level of individual countries in acknowledging its importance. These inadequate response mechanisms will result in the undermining of online GBV. It signals to a country's citizens that crimes committed online are not "serious" crimes.[60] That is far from the case; often, a victim's mental, psychological and even physical health are adversely affected. Thus, without adequate responsive measures, victims' and survivors' plight will continue to be undermined.

Conclusion

For ASEAN and its member states, to respond to the ever-increasing number of online GBV cases, an awareness needs to be built amongst the countries and their respective decision-makers. The region needs to see GBV as a matter that is seriously affecting the common interest of ASEAN while attempting to transcend the present challenge of the AMS' differing levels of digital infrastructure, cyber capability and acceptance of gender diversity's acceptance may arise as impediments. Coordinated and equal effort is necessary to facilitate the implementation of the Three Pillars Programme used to gauge for the inclusivity of cybersecurity frameworks. The Design Pillar must see the development of international, regional and national standardisations that are "gender-sensitive" and "gender-inclusive". The Defence Pillar should aim to have governments strengthen the internal and external regulation to further protect women from online GBV and encourage participation in the cyberspace. Finally, the Response Pillar raised the urgency for the AMS to start evaluating their existing response protocols towards victims and survivors, revising laws accordingly.

The AMS still face challenges in improving women's participation in the workforce, in technology sector and in parliaments, as well as breaking the conservativeness in translating gender to mostly women's issue. These hurdles need to be addressed to create a gender-sensitive approach to cybersecurity's Design, Defence and Response in the region, and addressing these challenges requires commitment, action and resources. Therefore, it is recommended that ASEAN strive to create a regional action plan to address online GBV. This can be done by taking the example of other regional organisation, such as the EU Gender Action Plan that include goals and actions for gender equality and women's empowerment, both online and offline. Admittedly, there are impediments that ASEAN states need to overcome prior to adopting a regional action plan to address online GBV. These include different perspectives and differing levels of urgency when it comes to addressing GBV, as well as different efforts and willingness to push for the articulation of existing regional agreements that were previously applied in the physical realm, such as the ASEAN Declaration on the Elimination of Violence Against Women (2004), to also apply in the digital spaces. Nevertheless, the borderless nature of cyberspace has called upon countries to take multilateral efforts in combating cybercrime. Regional efforts on prevention and response to online GBV are best implemented due to shared concerns and trends of cybercrime. Especially with the potential economic benefit cyberspace bring to ASEAN, ensuring people's safety on the internet should become a priority for ASEAN.

Notes

1 That is 400 million out of 580 million individuals. See, Google, Temasek Holdings and Brain & Company, *e-Conomy SEA 2020 Report* (2020) https://storage.googleapis.com/gweb-economy-sea.appspot.com/assets/pdf/e-Conomy_SEA_2020_Report.pdf

2 Ibid.

3 Office of the United Nations High Commissioner for Human Rights, *Report of the Special Rapporteur on Violence against Women, Its Causes and Consequences on Online Violence against Women and Girls from a Human Rights Perspective*, UN Document No. A/HRC/38/47, June 14, 2018.

4 UNFPA, *Covid-19 and Violence against Women: The Evidence Behind the Talk* (Bangkok: UNFPA, 2020), https://asiapacific.unfpa.org/sites/default/files/pub-pdf/covid-19_and_vaw_insights_from_big_data_analysis_final.pdf

5 Ibid.

6 ASEAN, *ASEAN Digital Masterplan 2025* (Jakarta: ASEAN Secretariat, 2021), p. 43.

7 Khanisa, "A Secure Connection: Finding the Form of ASEAN Cyber Security Cooperation", *Journal of ASEAN Studies*, Vol. 1, No. 1 (2013), pp. 41–53, and Catriona Heinl, "Regional Cybersecurity: Moving Toward a Resilient ASEAN Cybersecurity Regime", *Asia Policy*, Vol. 18, July 1 (2014), pp. 131–159.

8 Katherine Millar, James Shires and Tatiana Tropina, *Gender Approaches to Cybersecurity: Design, Defense, and Response* (Geneva: United Nations Institute for Disarmament Research, 2021), pp. 20–22. https:// doi.org/10.37559/GEN/21/01

9 Shires, J., "Family Resemblance or Family Argument? Three Perspectives of Cybersecurity and Their Interaction", *St Antony's International Review*, Vol. 14, No. 3 (2019), pp. 18–36, https:// www.ingentaconnect.com/content/stair/stair/2019/00000015/00000001/art00003

10 Deborah Brown and Allison Pytlak, "Why Gender Matters in International Cyber Security", *Women's International League for Peace and Freedom and the Association for Progressive Communications,* First Edition (2020), p. 2, https://www.apc.org/en/pubs/why-gender-matters-international-cyber-security

11 Araba Sey, "Gender Digital Equality Across ASEAN," ERIA Discussion Paper Series no. 358 (2020), https://www.eria.org/uploads/media/discussion-papers/Gender-Digital-Equality-Across-ASEAN.pdf

12 UN Women, *Action Brief on Women, Peace & (Cyber) Security in Asia and the Pacific* (Bangkok: UNFPA, 2020), p. 2, June 9, 2021, https://asiapacific.unwomen.org/-/media/field%20office%20eseasia/docs/publications/2020/06/action%20brief%20%20wps%20%20cybersecurity16620final.pdf?la=en&vs=1656

13 Ibid.

14 Zarizana A. Aziz, "Online Violence against Women in Asia: A Multicountry Study," *UN Women* (2020), p. 26, https://asiapacific.unwomen.org/en/digital-library/publications/2020/12/online-violence-against-women-in-asia

15 UN Women, *Action Brief on Women*, p. 2.

16 Aziz, *Online Violence against Women in Asia*, p. 33.

17 Ibid.

18 Save the Children and Plan International, *Because We Matter: Addressing COVID-19 and Violence against Girls in Asia-Pacific* (Singapore and Bangkok: Save the Children and Plan International, 2020), p. 8, June 11, 2021, https://reliefweb.int/report/world/because-we-matter-addressing-covid-19-and-violence-against-girls-asia-pacific

19 Aziz, *Online Violence against Women in Asia*, p. 18.

20 Komisi Nasional Anti Kekerasan Terhadap Perempuan, "Pengesahan RUU Penghapusan Kekerasan Seksual: Wujud Hadirnya Negara dalam Pemenuhan Hak Perempuan Korban Kekerasan Seksual", Press Release, November 24, 2020, https://komnasperempuan.go.id/siaran-pers-detail/siaran-pers-komnas-perempuan-peluncuran-kampanye-16-hari-anti-kekerasan-terhadap-perempuan-2020-24-november-2020

21 Aziz, *Online Violence against Women in Asia*, p. 26.

22 Social media companies have different speed and mechanism in countering to online GBV on their platforms. Meta, previously Facebook, has instilled "Community Standards" in 2019 to make its platform a safer place, while Twitter since 2020 has shown notification for people searching GBV keywords in its handlebar, containing the information on GBV as well as local organisations that offer help to address it. Meanwhile, TikTok only in 2021 partnered with UN Women to increase sensitivity of GBV. At the time this chapter was being written, majority of these social media campaigns were using English, making access limited to English-speaking Southeast Asians accessing the platforms.

23 Julie Posetti, Diana Maynard and Kalina Bontcheva, *Maria Ressa: Fighting an Onslaught of Online Violence*, International Center for Journalists (Sheffield: ICFJ, 2021), pp. 12, 31, and Safenet, "The Rise and Challenges of Doxing in Indonesia", June 29, 2021, https://safenet.or.id/2021/06/the-rise-and-challenges-of-doxing-in-indonesia/

24 Julie Posetti, Nermine Aboulez, Kalina Bontcheva, Jackie Harrison and Silvio Waisbord, *Online Violence against Women Journalists: A Global Snapshot of Incidence and Impacts* (UNESCO, 2020), p. 4, and Women's International League for Peace and Freedom (WILPF) and Association for Progressive Communication (APC), *Why Gender Matters in International Cyber Security* (New York and Johannesburg: WILPF and APC, 2020), pp. 13–14.

25 Ibid., p. 24.

26 Hein Thant Swe, "Combatting Cyberbullying in Southeast Asia: How ASEAN Can Adopt a Regional Approach", *ASEAN-Australia Strategic Youth Partnership*, August 26 (2019), https://aasyp.org/2019/08/26/combating-cyberbullying-in-southeast-asia-how-asean-can-adopt-a-regional-approach/

27 Mochamad Iqbal Jatmiko, Muhammad Syukron and Yesi Mekarsari, "Covid-19, Harassment and Social Media: A Study of Gender-Based Violence Facilitated by Technology during the Pandemic", *Pandemic in Society and Media,* Vol. 4, No. 2 (2020), p. 337, https://doi.org/10.26740/jsm.v4n2.p319-347

28 Ibid.

29 Nobuhle J. Dlamini, "Gender-Based Violence, Twin Pandemic to COVID-19", *Critical Sociology*, Vol. 47, No. 4–5 (July 2021), p. 585. https://doi.org/10.1177/0896920520975465
30 Nicolas Suzor, Molly Dragiewicz, Bridget Harris, Rosalie Gillett, Jean Burgess and Tess Van Geelen. "Human Rights by Design: The Responsibilities of Social Media Platforms to Address Gender-Based Violence Online", *Policy and Internet*, Vol. 11, No. 1 (March 2019), pp. 84–103, https://doi.org/10.1002/poi3.185, and Ruth Lewis, Michael Rowe and Clare Wiper, "Online Abuse of Feminists as an Emerging Form of Violence Against Women and Girl", *The British Journal of Criminology*, Vol. 57, No. 6 (November 2017), pp. 1462–81, https://doi.org/10.1093/bjc/azw073
31 UN Women, *Action Brief on Women*.
32 Ibid., p. 2.
33 Hoang Linh Dang, "Social Media, Fake News, and the COVID-19 Pandemic: Sketching the Case of Southeast Asia", *Austrian Journal of South-East Asian Studies*, Vol. 14, No. 1 (June 2021), pp. 37–58. https://doi.org/10.14764/10.ASEAS-0054
34 Ibid.
35 UNECE, "Gender Responsive Standards Initiative", *United Nations Economic Commission for Europe*, June 11, 2021, https://unece.org/gender-responsive-standards-initiative
36 Ibid.
37 Including ASEAN Declaration to Prevent and Combat Cybercrime (2017) and ASEAN Plan of Action in Combating Transnational Crime (2016–2025).
38 Srirath Goi Gohwong, "The State of the Art of Cybersecurity Law in ASEAN", *International Journal of Crime, Law and Social Issues*, Vol. 6, No. 2 (2019), pp. 12–23, http://dx.doi.org/10.2139/ssrn.3546333
39 Jay Cohen, Pichrotanak Bunthan, Dino Santaniello, Nwe Oo, Athistha (Nop) Chitranukroh, Gvavalin Mahakunkitchareon, Thao Thu Bui and Waewpen Piemwichai, *Regional Guide to Cybersecurity and Data Protection in Mainland Southeast Asia* (Bangkok: Tilleke & Gibbins, 2020), pp. 2–20, https://www.tilleke.com/wp-content/uploads/2020/07/Tilleke-Regional-Guide-to-Cybersecurity-and-Data-Protection-in-Mainland-Southeast-Asia.pdf
40 Deloitte, *Deloitte Asia Pacific Privacy Guide 2020–2021* (Singapore: Deloitte, 2021), https://www2.deloitte.com/id/en/pages/risk/articles/ap-privacy-guide-2020-2021.html
41 Siti Khairunnissa, Abdul Rahman Maulana Siregar and Andry Syafrizal Tanjung, "Law on Cyberbullying in Indonesia, Malaysia, and Brunei Darusallam", *International Conference of Asean Perspective and Policy (ICAP)*, Vol. 1, No. 1 (2018), pp. 63–69, https://jurnal.pancabudi.ac.id/index.php/ICAP/article/view/269
42 Lyndsay Freeman, "The New War on Women: Weaponising Online Spaces", Broad-Agenda, Oct. 29, 2020, https://www.broadagenda.com.au/2020/the-new-war-on-women-weaponising-online-spaces/
43 Ibid.
44 Vaishali Rastogi, Michael Meyer, Michael Tan and Justine Tasiaux, *Boosting Women in Technology in Southeast Asia* (Singapore: Boston Consulting Group, 2020), pp. 4–7, https://web-assets.bcg.com/6f/8c/a3914ef6482c910d76e00c2efde2/bcg-boosting-women-in-technology-in-southeast-asia-oct-2020.pdf
45 Ibid.
46 Ibid.
47 Nikkei Asia, "Southeast Asian Tech Hubs Race to Become the Next Silicon Valley", *Nikkei Asia*, Oct. 24, 2018, https://asia.nikkei.com/Spotlight/The-Big-Story/Southeast-Asian-tech-hubs-race-to-become-the-next-Silicon-Valley
48 Cristina Martinez-Fernandez and Marcus Powell, "Employment and Skill Strategies in Southeast Asia: Setting the Scene", *OECD Local Economic and Employment Development (LEED) Working Papers* (Paris: OECD Publishing, 2010/01), p. 59, http://dx.doi.org/10.1787/5kmbjglh34r5-en

49 The Asia Foundation, *The Future of Work Across ASEAN: Policy Prerequisites for the Fourth Industrial Revolution – Regional Summary and Recommendations* (Bangkok: The Asia Foundation, 2020).

50 Ingrid Brudvig, Nanjira Sambuli and Dhanaraj Thakur, "#eSkills4Policymakers: From Policy
Recommendations to Policy Action – Training Policymakers on Gender Equality in
ICT Policy Formulation", *Digital Skills Insights 2020 – International Telecommunication Union* (2020), pp. 71–80.

51 Araba Sey, "Gender Digital Equality Across ASEAN", *ERIA Discussion Paper Series*, no. 358 (2021), https://www.eria.org/uploads/media/discussion-papers/Gender-Digital-Equality-Across-ASEAN.pdf

52 IPU, "Women in Politics: 2021", *Inter-Parliamentary Union* (Geneva: IPU, 2021), https://www.ipu.org/women-in-politics-2021

53 ASEAN Working Group on Sustainable Development Goals Indicators, *ASEAN Sustainable Development Goals Indicators Baseline Report 2020* (Jakarta: ASEAN Secretariat, 2020), p. 60.

54 UN Women, *Action Brief on Women*, p. 3.

55 See, Ritchie, Roser, Mispy, Ortiz-Ospina. "Measuring Progress towards the Sustainable Development Goals: Sustainable Development Tracker Goal 5 Gender Equality and Empower All Women and Girls", *SDG-Tracker.org, website*, February 6, 2022, https://sdg-tracker.org/gender-equality

56 Save the Children and Plan International, *Because We Matter*, p. 9.

57 ASEAN Secretariat, *ASEAN Regional Plan of Action on the Elimination of Violence against Women* (Jakarta: ASEAN Secretariat, 2016), https://www.asean.org/storage/2012/05/Final-ASEAN-RPA-on-EVAW-IJP-11.02.2016-as-input-ASEC.pdf

58 UN Women, "Online and ICT-Facilitated Violence against Women and Girls during COVID-19", *EVAW COVID-19 Briefs* (Bangkok: UN Women, 2020), p. 3, https://www.unwomen.org/en/digital-library/publications/2020/04/brief-online-and-ict-facilitated-violence-against-women-and-girls-during-covid-19

59 Ibid.

60 Office of the United Nations High Commissioner for Human Rights, UN Document No. A/HRC/38/47, point 85.

5

ONLINE HARASSMENT AND THE SPACE FOR POLITICAL SPEECH

Advocating for Gender Equality in Malaysia

Gulizar Haciyakupoglu

Introduction

Malaysia's Ministry of Women, Family and Community Development offered various recommendations to women to avoid conflict at home during the pandemic-triggered lockdown, including advice to "avoid nagging", put on make-up, dress well and imitate the manga character Doraemon's voice when engaging with their husbands.[1] The sexist comments offered on social media sparked backlash.[2] They were later taken down and followed by an apology.[3] The case illustrated the imposition of gender roles onto women by an official body and social media's aid in spreading as well as countering gendered narratives.

Gender equality advocacy is paramount to the advancement of the gender equality agenda, and the Internet has been providing a platform for gender equality advocates to connect and drive their causes and for victims of gendered harassment and discrimination to access help. However, the same space also encapsulates online threats, such as harassment, that create security risks for gender equality advocates and expand the forms and avenues for violence against women (VAW). Indeed, various studies discuss how women are adversely affected by online harassment.[4] Threats, including harassment, in cyberspace impinge on the Internet's capacity to serve as a safe ground for gender equality advocates and VAW victims to vocalise their concerns, contribute to the definition of online security at a higher level and push for change. In this chapter, I explore this tension through the contentions surrounding what online security might mean to gender[5] equality advocates, victims, the government and other authorities in Malaysia, where the equality of the playfield for political speech has been debated,[6] and where the Internet provides opportunities for, and poses limitations to online gender equality advocacy concurrently.

DOI: 10.4324/9781003261605-7

I ponder how limitations to the space for online political speech may create insecurities ("lack of safety or protection"[7]) for gender equality advocates (and advocacy), especially when online harassment sparks security concerns and exacerbates VAW. Political speech, according to Political Studies Association, "concern[s] decisions about possible courses of action which are contentious and contested and about which people might reasonably disagree",[8] and, in this chapter, I interpret the speeches on gender equality and VAW-related issues by gender equality advocates as political speech. Some identity-based (including gender-based) digital abuse can be "understood as a struggle to control political discourse that reflects and reinforces existing social inequalities", as Sarah Sobieraj (2020) argues.[9] Sobieraj suggests that women who identify with various marginalised groups, who publicly share their views concerning (or in) masculine spheres and women who seen as feminists or as diverging from "traditional gender norms" are particularly impacted by digital attacks.[10] While her study focuses on a different geography, most of the cases shared in this chapter somewhat relate to these categories. More importantly, I believe gender equality advocacy is fundamental to advancing the gender equality agenda and countering VAW, including online harassment, and I see political speech as central to this endeavour. As such, impediments to political speech concerning gender equality advocacy and VAW risk undermining advocacy efforts and improvements to women's security.

While exploring the interaction between political speech, gender equality advocacy and online harassment in Malaysia, I consider how limitations to online expressions with the narrative justification to protect perceived social cohesion and "Asian values" might affect gender equality advocacy. My interest in this question was sparked after reading M. A. M. Sani's (2008) and S. Saleem's (2018) articles.[11] Sani (2008) argues that "opposition parties, associations and cause oriented groups['] [...] rights to political speech and their capacity to mobilise masses to impact on policy-making have been diminished" and safeguarding public security, sustaining Asian values and countering racisms are among justifications put forward when restricting political speech in Malaysia.[12] Saleem (2018) discusses the government's "securitisation of non-mainstream Muslims in Malaysia" citing public order and social cohesion-related concerns,[13] which may impact some of the advocacy efforts undertaken by organisations like Sisters in Islam (SIS). Citing Chew (1994), Mayer (1994) and Li (1996), she suggests that states have previously attempted to leverage "Asian values and Islamic civilisational approaches to human rights" as justifications to limit civil liberties.[14] Others also speak of governments offering national security and counter-terrorism as reasons to defend their restrictions on the freedom of speech and other rights.[15] In this chapter, discussions concerning social cohesion and "Asian values" narratives, and other limitations to political speech, are not limited to those coming from the government. However, the government is one of the main authorities that need to act to improve conditions.

The regulatory framework that draws boundaries around political speech, and legal protections against online harassment, influence the security and freedom of online spaces for advocacy and discussions surrounding gender equality and VAW-related concerns. Hence, in addition to narratives on social cohesion and Asian values, I explore the insecurities that may arise from the presence or absence of regulations concerning online harassment and political speech. The discussions in this chapter lead me to identify two immediate needs with regards to the topic: the negotiation of what online harassment constitutes for different stakeholders in Malaysia,[16] and the expansion and security of the space for civil and inclusive political speech.

This chapter complements arguments from relevant literature with insights from interviews I conducted for my doctoral dissertation on the *Internet and gender equality advocacy within the Islamic context of Malaysia.*[17] The dissertation focused on a broader topic, and I will only refer to interviewee accounts on online intolerance[18] in this chapter. As such, these accounts should be taken as anecdotal examples rather than a systematic analysis based on VAW-specific questions. Here, I also acknowledge that there have been social and political changes in Malaysia since I conducted the interviews and since the publication of Sani's, Saleem's and some other prominent articles on intersecting topics. For a better understanding of events on the ground, I complement the scholarship above with data and information from contemporary studies published by organisations such as United Nations (UN) Women.

Following this introduction, the next section will set the context on VAW in online spaces. The section will help establish the relationship between gender equality advocacy, including efforts on VAW, and political speech. Subsequently, the "Online security and political speech" section will consider online security conceptualisations in relation to the debates surrounding "social cohesion", "Asian values" and regulation of political speech, with attention to VAW and gender equality advocacy. The chapter will conclude with a discussion of policy considerations.

VAW in cyberspace and gender equality advocacy in Malaysia

The Convention on the Elimination of All Forms of Discrimination against Women (CEDAW) identifies gender-based violence as "violence that is directed against a woman because she is a woman or that affects women disproportionately".[19] VAW materialises in multiple forms, including domestic violence, sexual violence and harassment, and privacy violations,[20] some of which occur online or are facilitated by technology and cyberspace. Correspondingly, the European Parliament's Policy Department for Citizens' Rights and Constitutional Affairs contextualises "cyber violence and hate speech online against women", including cyber harassment, within gender-based violence[21] (GBV).[22] Association of Southeast Asian Nations (ASEAN) also recognises "online harassment, abuse, bullying, stalking and distribution of denigrating images" as types of VAW.[23]

Amidst the multiplicity of terms, many forms of cyber violence "remain under-defined",[24] leading bodies like UN Women to call for "clarity and consensus among States, ICT intermediaries and civil society on what constitutes online violence against women".[25] According to a 2020 UN Women report, culture and society influence what is viewed as information and communication technologies (ICT)-facilitated violence against women and girls (VAWG), and states gravitate towards seeing "violence as a criminal offence", prioritising "grievous physical and sexual harm, and trivializing non-grievous harm".[26] There is a risk of VAW-related issues being subsumed under broader topics (e.g., public safety) in awareness-raising initiatives in media or downgraded to a women's concern,[27] especially when the voices of gender equality advocates and victims are missing from the discussion. This could lead to a significant societal threat being underrated,[28] amplify the security concerns of women and vulnerable groups,[29] and result in the misallocation of importance and responsibility in the solutioning process when a whole of society approach is necessary.

Arriving at a shared understanding of online and ICT-facilitated VAW is a fundamental step in improving communication and policymaking on the issue.[30] Here, how different definitions would be negotiated and which stakeholders' views would be considered when forging shared understandings emerge as questions. These questions have particular relevance as victims and activists have their own views on what security from online harassment mean, and government and other authorities – arguably – approach the issue from their political stance. Besides, social media platforms need to be more transparent on the extent and types of gendered harms happening on their platforms to improve understanding on the issue.

Against the expanse of definitions and the expanding threat landscape, gender equality advocacy has been carving a space for itself in Malaysia. Advocacy efforts on VAW have featured success. Some argue that the "most acknowledged forms of VAW" are those that women's groups have vigorously advocated for, including "domestic violence, rape and sexual harassment".[31] The mid-1990s was an important period, with the introduction of the Domestic Violence Act (DVA) and the ratification of the CEDAW.[32] To some, overall attention on VAW changed in scope over time.[33] The issues surrounding women's role in the economy and family, and their reproductive roles came into prominence, risking an eclipse of broader and intersecting issues concerning women.[34] A myopic view of VAW and gender equality may obscure the extent of the issue, and advances in select areas may improve conditions to a degree but could fall short in fully resolving the multifaced problem. For instance, Malaysia amended the Employment Act (1955) to address sexual harassment better,[35] and some criminal laws cover sexual harassment related offences.[36] However, victims may not be in a position to afford the cost of a lawsuit, those in charge may not see some sexual harassment cases as criminal and existing laws alone are not adequate to eradicate the problem of online sexual harassment.[37] The DVA has also been amended since its introduction, but its implementation can be enhanced.[38]

Despite positive developments, VAW remains a fundamental issue, as signalled by the surge in domestic violence cases during the pandemic in Malaysia.[39] Besides, women's sexualisation in media continues to be a concern,[40] and it may exacerbate the problem by contributing to the normalisation of harassment.[41] Gender equality advocacy can help push for a change, and the Internet can provide a ground for that – but to a certain extent and not without impediments. Malaysia's Multimedia Super Corridor (MSC) Bill of Guarantees, which promised no censorship online, raised hopes for the expansion of the space for free speech,[42] including speech by "opposition and civil society movements".[43] The Internet opened up spaces for the opposition, alternative opinions and advocacy[44] – to a certain extent. Various interviewees I have spoken to suggested that the Internet has provided an arena to express and access alternative views,[45] some of which are not available or cannot be shared as freely offline, and some said Internet affords a platform for:

- Speaking bolder on challenging faith-related issues,[46]
- Advocating gender equality,[47]
- LGBT[IQ+] and marginalised groups to speak and receive support,[48]
- Participating in discussion without the need of physical travel,[49]
- Globalising conversations with the possibility of people around the world accessing them,[50]
- Respite in a context where laws limit freedom of speech[51] and
- Engaging with diverse opinions,[52] including an interaction between conservative and progressive opinion holders.[53]

Besides, an interviewee argued that closed Facebook groups provide a safe space for women to engage.[54]

Despite the opportunities these interviewees listed, there is room for improvement and women still face impediments in accessing political speech and decision-making. For instance, the number of women in politics and, obliquely, women's access to political decision-making remain limited,[55] with Malaysia ranking 128 out of 155 countries in political empowerment in World Economic Forum's 2021 Global Gender Gap Report.[56] Gender equality advocacy can bring about change. However, online threats (Section "Online harassment and security") and limitations to political speech (Section "Online security and political speech"), among others, risk undermining the Internet's capacity to aid gender equality advocacy, curtailing their space for advocacy and creating insecurities for gender equality advocates and VAW victims. I will discuss them below.

Online harassment and security

A female user shared a teacher's disturbing comments on rape on TikTok and kindled a conversation about harassment in schools.[57] Social media allowed her to spread her message and garner support, but it was also used to send her hate messages.[58] The incident showed how standing up against abuse online may bear benefits and risks concurrently and how online and offline spaces are enmeshed, precluding a separation between online and offline security. The means provided by the ICTs can both be used to advance gender equality advocacy and commit VAW.[59] Examining the concerns of women, gender equality advocates, in particular, could provide a window into what they see as security issues against the question of an institutional approach to online security. This is vital as a definition of online harassment from a security lens that does not attend to their concerns might exacerbate their insecurities.[60] Incorporating their views would inform endeavours to cultivate good uses of ICTs and allow authorities to address security from harassment with attention to gender-based experiences.

The types of online harassment against women in Malaysia include privacy invasion, death or rape threats, sexual harassment and hate speech,[61] and some online gender-based violence couch themselves in "ethno-religious framework".[62] Civil Society Organisations (CSOs) and activists seek to help victims of and counter online harassment, while they also face harassment in online spaces.[63] In a UN Women (2020) report on "online violence against women in Asia", CSOs in Malaysia voiced facing attacks due to their "work and political stances".[64] According to the report, "Muslim women's advocates" were subject to online harassment over "dress and behaviour", with examples including harassment against a female politician and a celebrity for how they wear hijabs and another female politician for not wearing make-up and for backing a CSO working on Muslim women's right, which the attackers "deemed 'deviant'".[65] Other reports also mention religious views framing some of the responses to women's online expressions,[66] at times "impos[ing]" an *interpretation of* "religious morality".[67]

In some of my interviews, study participants also shared their observations and experiences on intolerances, some of which had gendered undertones or were against activists or alternative opinions.[68] One interviewee, who was then an employee of an NGO working on gender issues, mentioned an incident where a woman attacked a member via "fake stories".[69] One shared how some women are "intimidated talking offline", and some attacks target women identifying as feminists.[70] Another argued that some "cybertroopers" attack "the person, family, mother …" rather than the issue and recounted a case where a Malay Muslim person "announc[ing] he is gay" faced attacks that led to him deactivate the video and hide.[71] Two interviewees noted how anonymity could be abused when misbehaving online,[72] raising a question on the relation between the responsibility of behaviour and anonymity. On personal experiences, an interviewee shared how her comment suggesting that "there are different interpretations on hijab" was "bombarded" and argued that "some rigid religious groups […] [that] think there is one kind of Islam" may attack "when there is a different view".[73] Another interviewee suffered from an online attack where she was wrongly accused of writing a comment that was found to be insulting to Islam.[74] She received threats until it became clear that she was not the entry's author.[75] And there were interviewees who commented on values and rules of discussion groups, organisations' responsibility in managing online discussions or cited case(s) where online harassment on an organisation's Facebook page led to a harasser being banned.[76] These are all different examples, but each involves threats and challenges the space for civil, free and secure political speech.

Overall, the accounts of online harassment exhibit the permeation of persevering problems such as misogyny and gender inequality in online spaces and ICT's facilitation of certain threats. The examples encapsulate attempts to oppress women's agency and claim over their own bodies, hinting at the problem cutting across physical and virtual boundaries – if there is any. Gender-based online harassment could create a fear of engagement on certain issues in online spaces,[77] lead some to disconnect from an online platform,[78] further ingrain existing biases and inequalities, inhibit civil engagement between bearers of differing opinions and take a toll on social cohesion. Furthermore, online harassment, in some cases, may deepen fault lines that domestic and foreign malicious actors might exploit,[79] and with that not only pose a threat to individuals, but also have social implications and constitute national security concerns. For advocacy efforts to get heard, it is necessary to have a platform that is free from restrictions on speech concerning gender equality advocacy and secure from harassment, where security is defined with respect to differences in experience and opinions. Authorities have to take victims' and gender equality advocates' views on online harassment into account when exploring what security from online harassment mean and evaluate policies against the risk of them creating insecurities for victims and gender equality advocates.

Online security[80] and political speech

In this section, I ponder how limitations to political speech may create insecurities for the gender equality advocates and victims of VAW and impede efforts to curb harassment. First, I contemplate how the use of "Asian values" as a set of narratives to justify the control of political speech might lead to the downplaying of women's security concerns and advocacy initiatives (Section "Asian values and feminism"). Second, I consider how the authorities' approach to "social cohesion", and the interaction between religion and gender might influence gender-based insecurities (Section "Social cohesion and insecurities"). Lastly, I discuss how the limitations and protections afforded by the legal framework might affect online gender equality advocacy (Section "Regulation of online political speech"). The variance in approaches to security, social cohesion, online harassment and Asian values between government (and other authorities such as religious authorities) and advocacy groups (and VAW victims) lurks behind the discussions in each section. The sections below involve theoretical considerations that have not necessarily materialised, yet the risk of their possible occurrence necessitates reckoning with these concern.

Asian values and feminism

In Malaysia, "Asian values", which "stresses the role of the culture, including religion, in determining the identity and distinctiveness of the Asian peoples"[81], are among the justifications offered to control political speech according to some scholars.[82] Besides, as also mentioned above, values, including those related to religion, could be used in framing the responses to women's online expressions. "Asian values" is occasionally "compar[ed] and contrast[ed]" with "certain Western values" when underlining the "definitive elements of 'Asian democracy'",[83] despite the vagueness of both terms and the difficulty of subsuming various cultures within those geographies under a single label. Feminism[84] occasionally comes under the spotlight in value-oriented discussions, and, in and beyond Malaysia, some view feminism as Western or un-Islamic.[85] The labelling of "feminism" as a "Western" concept and its juxtaposition against "Islam" emerged as a topic during some of my interviews. Various interviewees shared how some see feminism as "un-Islamic",[86] and some interviewees mentioned the interpretation of feminism – and Islamic feminism – as a Western ideology[87]. Among them, one argued that some policy and lawmakers think "gender equality is something that is foreign so if they [women] say they are feminist, they open themselves to attacks … like [they are] not Islamic enough …".[88]

As per this chapter, framing feminism as a foreign, "Western" or "un-Islamic" concept that is against Asian values creates several risks: hesitancy to embrace a gender focus when evaluating crucial problems; side-lining of gender equality advocacy efforts; making gender equality advocates (regardless of whether they identify as feminists) vulnerable to online attacks and falsely casting them as a group disconnected from issues on the ground. The latter would be an ignorance of the variety of views within feminism, and how feminism in Malaysia has been interacting with the economic, social and political changes in the country. For instance, according to Mohamad et al. (2006), "distinctive historical moments have engendered many forms of feminist politics and practices".[89] Mohamad et al. acknowledges the possible overlap between different strands of feminism across periods while stating that "nationalist feminism, social feminism, political feminism and market feminism have coincided as markers of the country's transformation from colonialism to a post-Reformasi state".[90] Hence, Malaysian feminism and the women's movement have not been disconnected from the country's political context. They adapted to the political, social and economic changes and responded to the new conditions, inequalities and gaps created by these shifts.

As the interplay between the political context and feminism continues, global-local views on gender issues interact and clashes in values arise concurrently. On the one hand, global movements such as #Metoo find a reflection in different localities, local advocacies garner international support and intersections between local and global concerns allow movements to gain a voice. On the other hand, resistance against global movements emerge, with values occasionally put forward as justifications for restrictive actions, at times on issues concerning gender equality. In this context, although the examples above did not specifically focus on government, the government's approach to feminism and how it situates gender equality in a relation to Asian values could tilt the dynamics and influence public and political perspective on both. While time will reveal more on the current and future governments' outlooks, discussions in this section urge the consideration of the cost to gender equality advocacy efforts if feminism and gender equality advocacy are portrayed as in conflict with Asian values.

Social cohesion and insecurities

The progress towards gender equality and tackling VAW could strengthen social cohesion by contributing to the establishment of a more equal and inclusive society. This is one of the reasons why authorities and the public should care about the fight for gender equality and against VAW. However, the complexities arising from the interaction between authorities' view of security threats and "social cohesion" and that of gender equality advocates may find body in issues that limit space for advocacy and create insecurities for women in Malaysia.

Securitising an issue involves the framing of a "societal insecurity [...] as an existential threat",[91] and it may result in a group being perceived to "pos[e] a societal threat".[92] In some cases, this group could involve gender equality advocates and VAW victims. Similarly, "othering" gender equality advocates when their fight concerns the society at large, disagreements between authorities and advocacy groups on VAW-related issues, and lack of adequate protection against threats faced by gender equality advocates may cause cracks in social cohesion and erupt security concerns for gender equality advocates. Besides, the securitisation attempt may involve the curtailment of their access to political speech and render VAW victims and gender equality advocates insecure.

A case concerning SIS reflects some of the concerns above, demonstrates the possible clashes between religious, political and gender-oriented approaches, and renders the space available for the voices of those advocating for gender equality vulnerable to scrutiny. In 2014, state of Selangor's fatwa "declared SIS to be deviant due to its alleged liberalism and religious pluralism beliefs"[93] and requested "Malaysian Communications and Multimedia Commission, to monitor and block social media websites" that had liberalism and religious pluralism related content, which they believed was "against Islam".[94] Based on a review of "news reports and public statements by the state on [...] religious deviancy restrictions on the Shia minority group and [...] Sisters in Islam under the Barisan Nasional government between 2010 and 2015", Saleem (2018) argued that the "elected politicians" and religious bureaucracy of the time leveraged "the public order and social cohesion" claims as justifications in their "securitising discourse".[95] While this case may come across as one that concerns SIS only, gender and religion interact in many cases in Malaysia and organisations aiding women and the LGBTIQ+ community need to be informed on the Islamic context in Malaysia to offer context-sensitive help where necessary.[96]

When balancing between social cohesion and security, authorities need to consider gender equality as a factor influencing this balance and see advancement of gender equality and tackling of gendered impacts of security concerns as steps that could contribute to both social cohesion and security.

Regulation of online political speech

The vocalisation of gender equality advocacy efforts, and concerns on VAW, is essential for advancement in these areas. However, how authorities use the legal[97] framework to regulate the space for speech may influence gender equality advocates' room to drive their cause and create insecurities for them. On the other hand, absence of or problems in the enforcement of inclusive and carefully drafted laws concerning online harms may render gender equality advocates and women less secure in online spaces.

Various experts[98] cite cases of or suggest government control over content via laws sometimes with the justification of safeguarding political and racial harmony,[99] or national security,[100] and some[101] note the use of such laws to regulate online acts and content, despite the no-censorship promise of the MSC. This is important as restrictive approaches may encumber oppositions and "cause oriented groups'" access to "political speech and their capacity to mobilise masses to impact policy making".[102] While some issues concerning VAW and gender equality advocacy may be less contentious and would not face any restrictions, on issues where gender equality advocacy groups and the government (and other authorities) hold severely clashing views, advocacy efforts that are critical of authorities' actions may also face a risk of restrictions. Also, how laws are interpreted and enforced may discourage some from speaking up.

Above questions call for an observation of cases where gender equality advocates' views clash with authorities' stance. I should note that these considerations do not to mean that the available laws absolutely restrict or prevent people from mobilising, or that there is no space for political speech: as social media have facilitated advocacy in some cases[103] and the Internet has provided space for political speech, as also mentioned when discussing VAW advocacy above. However, the conditions for free and political speech still need improvement. Malaysia has been through changes in government in the last few years, but the degree of change in space for free speech remains questionable. A 2020 Human Rights Watch publication suggested that the previous administration under Prime Minister Muhyiddin Yassin "backslid[] on Free Speech", with the use of some "laws to investigate and prosecute speech critical of the government".[104] Also, some non-profit organisations suggest that while the GBV aspect of gender inequality is often "addressed", the efforts to mitigate the effects of "gender inequality in the access and exercise of freedom of opinion and expression" remain inadequate.[105]

Laws, depending on the fairness of their content, enforcement and other factors may help curb online harassment, and with that, obliquely help provide safer conditions for online advocacy. According to UN Women, "75% of legal frameworks that promote, enforce and monitor gender equality under the SDG indicator, with a focus on violence against women, are in place" in Malaysia.[106] The efforts to improve this percentage should continue while the gender-sensitive enforcement of available laws have to be continuously monitored. There is also a need to better address intersecting areas where there is insufficient gender-focused protection against various forms of online harassment. According to a 2020 UN Women report, Malaysia has cybercrime courts, but "they do not seem to address ICT VAWG",[107] suggesting the need to introduce a gender-focus to existing institutions and practices. Besides, some laws in place may not be bullet proof, and some can be interpreted as discriminatory or unequal on the basis of gender-identity and sexual orientation.[108] Also, existing laws, such as the Communications and Multimedia Act (1998) Section 233, may not always be enforced in cases of online GBV.[109] A review of access to and enforcement of laws with a gender-focus would reveal the extent of necessary revisions and improvements.

In 2021, in his keynote speech to ASEAN Digital Ministers' Meeting (ADGMIN1), Prime Minister of the time, Sri Muhyiddin Yassin, called for legislation against hate speech on digital platforms, and said: "This is not limited to harassment and threats against a person or persons based on their race, gender, religion, sexual orientation, disability or nationality".[110] However, variance in approaches to and definitions of cyber harassment raise questions on coverage and inclusivity of the existing and future laws on the issue. Also, misogynistic and sexist remarks,[111] such as the one shared in the introduction, cast doubt on the degree of empathy, and attention that gender and intersectional concerns would receive in the drafting of any future laws and in the enforcement of existing laws on the topic. Moreover, the representation of women in the Civil and Sharia judiciary needs improvement.[112]

Corresponding to above discussions, some interviewees noted various limitations to political speech in their comments, which encapsulated legal restrictions and VAW-related issues and beyond. They included pressure faced when engaging in political reporting in Malaysia,[113] and the question of Internet's freedom from limitations or censorship,[114] with one raising increasing permeation of limitations to and the emergence of new ways to control online speech (e.g., jamming, cyberattacks, etc.),[115] and another seeing cyberspace as an alternative platform against the control of speech in some traditional channels of communications.[116] On censorship, some interviewee accounts highlighted the risk of self-censorship. How charges against some Twitter users create a chilling effect,[117] and the influence of community pressure over how people engage with sensitive content (e.g., publicly visible like versus sending a private message)[118] were among comments. As per the interviewee account mentioning how some prefer to communicate contentious issues via private messaging rather than publicly visible posts, exploring what remains private may help identify issues of tension, and cases where self-censorship takes hold. Lastly, some interviewees noted other factors that could influence the scope and reach of online content, including group/forum rules,[119] language preferences of audiences and advocacy groups and based on topic,[120] potential grouping of the like-minded[121] and algorithmic curation of newsfeeds.[122]

Policy considerations and concluding remarks

Previous sections demonstrated the need for different stakeholders to negotiate what online harassment means,[123] and for the expansion of the space for political speech specifically that pertaining to gender equality and advocacy against VAW. Arriving at an inclusive definition of online harassment (and other threats in cyberspace) that takes the views and experiences of various stakeholders into account requires studying the problem, and then negotiating different approaches. The expansion of the space for political speech, on the other hand, calls for a fundamental change that would address underlying problems, improve conditions for freedom of speech and institute civil, inclusive and respectful online engagement. Both are components of the broader task of responding to underlying, current and future problems concerning gender inequality and VAW.

Understanding and negotiating what online harassment constitutes

Scholars, civil society and intergovernmental organisations and the government have to expand studies exploring online threats from a gender-lens and with attention to the security landscape within Malaysia, and as others also argue, they need to gather gender-focused data. Increasing studies on the topic would help improve the understanding of the gendered implications of online threats and the curtailment of political speech; identify interpretational and contextual variances, and deracinate overlaps and differences between cases.[124] As Malaysia's political context has been dynamic in recent years, and there is need for up-to-date research on issues highlighted in this chapter, scholars, experts and other stakeholders need to observe and reflect on the changes and challenges closely. The ideal is to translate the learnings from studies into inclusive policies.

Furthermore, regulatory framework requires extra attention. Scholars, practitioners, intergovernmental organisations and CSOs working on intersecting issues can study the existing legal framework, including Syariah law, with a gender-focus to better understand its scope and record how existing laws are enforced, as this would help identify potential insecurities some laws and their enforcement might create for gender equality advocates and groups that are vulnerable to gender-based harassment.[125] There are already some relevant measures available, such as SIS's Telenisa, which offers "free legal advice service on [...] issues related to Shariah Islamic Family Law such as divorce, polygamy, alimony of wife and children, matrimonial property and others",[126] and provides statistics and reports on its findings.[127] Such efforts can be expanded in scope and volume. Other regulation-related considerations include taking action against "secondary perpetrators" who partake in VAWG by "downloading, forwarding and sharing VAWG content by third parties (principal perpetrators)", and providing "training for specialized law enforcement, prosecutors and judges on the specificities of ICT VAWG".[128] The legal frameworks should balance between national security and freedom of speech,[129] and while doing so, ensure that there is enough protection against online VAWG.

Inciting the interest of various stakeholders in these studies and integrating them in discussions and raising their awareness are vital in cultivating comprehensive research, inclusive definitions and resource investment. Increasing knowledge on these issues requires social media companies to be more transparent on the number of cases submitted for consideration and processed under violence and hate speech in Malaysia, and how they compare to those in other countries. This would also be a step towards building a bridge between what these companies observe on their platforms and what victims experience as users of the platforms. Local attempts to define and curb online violence and hate speech vis-a-vis international attempts is another area to explore, especially when populations, including victims of online harassment, have access to local and global discourses and platforms where online harassment and countermeasures take place.[130]

Expanding the space for political speech

The expansion of the space for civil, and inclusive speech, primarily requires a willingness to improve conditions for the freedom of speech while at the same time balancing freedom of speech with safe and civil engagement. Addressing root causes of ICT-facilitated VAWG,[131] and embracing a gender-focused approach to them is paramount to reach this balance and sustainable solutions. Raising public awareness is a part of addressing underlying concerns. It is especially important when there is a fragment of the society that believes enough has been achieved: in a 2019 IPSOS survey where "more people disagree[d] (49%) than agree[d] (42%) that when it comes to giving women equality, things have gone far enough" – those agreeing still constituted a significant percentage.[132] Furthermore, social media companies that host some of the advocacy efforts as well as online harms have to do more to protect the safety of their users.

Expansion of the space for civil and inclusive political speech – and tackling underlying issues – are long-term endeavours that demand social, political and institutional change.[133] For instance, according to a 2020 UN Women report, one of the main reasons CSO's think impede victims reporting ICT-facilitated VAWG is "lack of confidence in police",[134] which suggests a need to review practices, build confidence and configure reporting mechanism that victims find easy and feel safe to use. The latter includes both online and offline reporting mechanisms – not only to government, but also to the platforms. It is also essential to boost women's participation in decision-making *in all aspects of life*[135] as harassment knows no boundaries and solution involves multiple areas from industry to government, online to offline.

These and other measures that would contribute towards a more civil, and inclusive environment for political speech, could open up a safer space for the voices of advocates and victims, and using this space, they may push for gender equality and online safety. The voices of advocacy and civil society are fundamental to achieving change. They play an indispensable role in holding government, social media platforms and other stakeholders accountable on online harms, including gender-based harassment. When creating fundamental change in approaches to online threats, introducing a gender-focused view and establishing a safe and inclusive space for political speech is concerned, change has to happen in every aspect of life. And this primarily requires a comprehensive, relational mapping of interconnected and underlying problems, and circles back to the need to increase knowledge on the issue.

Acknowledgement

I would like to thank Assoc. Prof. Maznah Mohamad, Dr. Tamara Nair and Benjamin Ang for sharing their views on an earlier version of this chapter.

Notes

1 Low Zoey, "'Talk Like Doraemon': Malaysian Ministry Issues Tips for Wives during COVID-19 Movement Control Order", *Channel News Asia*, 31 March 2020, https://www.channelnewsasia.com/news/asia/coronavirus-malaysia-ministry-tips-wives-nagging-doraemon-mco-12593708; Kim Elsesser, "Malaysian Government Apologizes After Advising Women To Wear Make-Up And To Avoid Nagging During Lockdown", *Forbes*, 2 April 2020, https://www.forbes.com/sites/kimelsesser/2020/04/02/malaysian-government-apologizes-after-advising-women-to-wear-make-up-and-to-avoid-nagging-during-lockdown/?sh=6d4fdc8a3537
2 Ibid.
3 Ibid.
4 Sarah Sobieraj, "Credible Threat: Attacks against Women Online and the Future of Democracy", New York: Oxford University Press, 2020; Nina Jankowicz, Jillian Hunchak, Alexandra Pavliuc, Celia Davies, Shannon Pierson and Zoe Kaufmann, "Malign Creativity: How Gender, Sex, and Lies Are Weaponized against Women Online", The Wilson Centre, Science and Technology Innovation Program, January 2021, https://www.wilsoncenter.org/publication/malign-creativity-how-gender-sex-and-lies-are-weaponized-against-women-online; Julie Posetti, Nabeelah Shabbir, Diana Maynard, Kalina Bontcheva and Nermine Aboulex, "The Chilling: Global Trends in Online Violence against Women Journalists", UNESCO, April 2021, https://unesdoc.unesco.org/ark:/48223/pf0000377223; Cécile Guerin and Eisha Maharasingam-Shah, "Public Figures, Public Rage: Candidate Abuse on Social Media", Institute for Strategic Dialogue, 2020, https://www.isdglobal.org/wp-content/uploads/2020/10/Public-Figures-Public-Rage-4.pdf.
5 Despite references to gender, the reviewed literature and interview accounts focus more on women.
6 Mohd Azizuddin Mohd Sani, "Freedom of Speech and Democracy in Malaysia", Asian Journal of Political Science, 16:1, 85–104, 2008, DOI: 10.1080/02185370801962440
7 Merriam-Webster, https://www.merriam-webster.com/dictionary/insecurity.
8 "What makes a speech political?", Political Studies Association, https://www.psa.ac.uk/what-makes-speech-political.

9 Sobieraj, p. 3.

10 Ibid, p. 10.

11 Sani, 2008; Saleena Saleem, "State Use of Public Order and Social Cohesion Concerns in the Securitisation of Non-mainstream Muslims in Malaysia", Journal of Religious and Political Practice, 4:3, 314, 2018, https://doi.org/10.1080/20566093.2018.1525899.

12 Sani, 2008, p. 88.

13 Saleem, p. 314.

14 Ibid, p. 319.

15 Agnes Callamard, "Global Freedom of Expression and National Security: Balancing for Protection", An overview and analysis written for Columbia Global Freedom of Expression Judges' Training Materials, Global Freedom of Expression Columbia University, December 2015, https://globalfreedomofexpression.columbia.edu/wp-content/uploads/2016/01/A-Callamard-National-Security-and-FoE-Training.pdf.

16 Various experts highlighted the variance in approaches to online harassment: Nina Jankowicz et al., January 2021, p. 2, 38; Lucina Di Meco and Saskia Brechenmacher, "Tackling Online Abuse and Disinformation Targeting Women in Politics", Carnegie Endowment, 30 November 2020, https://carnegieendowment.org/2020/11/30/tackling-online-abuse-and-disinformation-targeting-women-in-politics-pub-83331; Kirsten Zeiter, Sandra Pepera, Molly Middlehurst and Derek Ruths, "Tweets That Chill: Analyzing Online Violence Against Women In Politics: Report of Case Study Research in Indonesia, Colombia, and Kenya", National Democratic Institute (NDI), 2019, p. 16, 17, https://www.ndi.org/sites/default/files/NDI%20Tweets%20That%20Chill%20Report.pdf.

17 I Interviewed 18 people fitting at least one of the following for my dissertation: "feminist, NGO employee or volunteer working within the Islamic framework with a progressive approach, communications professional or academic" (p. 392). Interviews shared here are selected from dissertation's Chapter 7 – noted if otherwise. Where applicable, letters in footnotes correspond to interviewee aliases. Gulizar Haciyakupoglu, "Feminism, Interpretation, Authority: A Critical Analysis of the Internet and Gender Equality in the Islamic Context of Malaysia", Dissertation (Unpublished) submitted to the National University of Singapore in 2016, https://scholarbank.nus.edu.sg/handle/10635/144370.

18 "Intolerances" encapsulates various degrees of negativity – used to avoid enforcing violence, harassment and alike onto interviewees. Dissertation's Chapter 7, among others, touches on intolerances.

19 "General Recommendations Adopted by the Committee on the Elimination of Discrimination Against women", Eleventh session 1992, General Recommendation No: 19: Violence against women, item 1 and 6, The Office of the High Commissioner for Human Rights, https://tbinternet.ohchr.org/Treaties/CEDAW/Shared%20Documents/1_Global/INT_CEDAW_GEC_3731_E.pdf.

20 Jac sm Kee and Sonia Randhawa, "Malaysia: Violence against Women and ICT", Association for Progressive Communications (APC), September 2009, pp. 12–23, https://www.genderit.org/sites/default/files/malaysia_APC_WNSP_MDG3_VAW_ICT_ctryrpt_0.pdf

21 VAW and GBV are not the same.

22 Adriane Van Der Wilk, "Cyber Violence and Hate Speech Online against Women: Women's Rights and Gender Equality", *European Parliament Policy Department for Citizens' Rights and Constitutional Affairs*, Directorate General for Internal Policies of the Union PE 604.979, September 2018, p. 11, https://www.europarl.europa.eu/RegData/etudes/STUD/2018/604979/IPOL_STU(2018)604979_EN.pdf.

23 "ASEAN Regional Plan of Action on the Elimination of Violence against Women (ASEAN RPA on EVAW)", The ASEAN Secretariat Jakarta, February 2016, p. 7, https://cil.nus.edu.sg/wp-content/uploads/2019/02/2016-2025-RPA-on-Elimination-of-Violence-against-Women-1.pdf.

24 Van Der Wilk, p. 13.
25 UN Women, "Online Violence against Women in Asia: A Multicounty Study", November 2020, p. 11, https://www2.unwomen.org/-/media/field%20office%20eseasia/docs/publications/2020/12/ap-ict-vawg-report-7dec20.pdf?la=en&vs=4251
26 Ibid, pp. 11–12.
27 Abigail de Vries in Kee and Randhawa, p. 6.
28 Kee and Randhawa argue that VAW issues do not receive the same sense of "urgency and politicisation as other rights-based issues such as the right to peaceful assembly", p. 6.
29 Endnote note 60; Rita Floyd, "Towards a Consequentialist Evaluation of Security: Bringing Together the Copenhagen and the Welsh Schools of Security Studies", Review of International Studies, 33: 2, 335, 336, 2007, https://www.jstor.org/stable/40072168.
30 Similarly, Sobieraj argues that having "a shared language around what it means to be targeted online will help advocates press for improved institutional responses that center victims" (p. 150). Similar calls to reach a shared understanding on terms are prevalent in influence operations literature.
31 Kee and Randhawa, p. 5.
32 "Human Rights Commission of Malaysia An Independent Report to the Committee on the Convention on the Elimination of All Forms of Discrimination against Women (CEDAW)", Human Rights Commission of Malaysia, 2017, item 2.1, p. 3, https://tbinternet.ohchr.org/Treaties/CEDAW/Shared%20Documents/MYS/INT_CEDAW_IFN_MYS_27118_E.pdf; Kee and Randhawa, pp. 8.
33 Kee and Randhawa, pp. 7–12.
34 Ibid.
35 "Human Rights Commission of Malaysia", item 6.7, p. 11.
36 "JAG: Yes, We Need the Sexual Harassment Bill", Women's Aid Organisation, 17 September 2021, https://wao.org.my/jag-yes-we-need-the-sexual-harassment-bill/.
37 Ibid.
38 For details: "The Status of Women's Human Rights: 24 Years of CEDAW in Malaysia", Women's Aid Organisation and Joint Action Group for Gender Equality", 2019, pp. 56–59, https://wao.org.my/wp-content/uploads/2019/01/The-Status-of-Womens-Human-Rights-24-Years-of-CEDAW-in-Malaysia.pdf.
39 Women's Aid Organisation (April 2020) in Quilt.AI, UN Women, Women Count and UNFPA, "Covid-19 and Violence against Women: The Evidence Behind the Talk: Insights from Big Data Analysis in Asian Countries", 3 March 2021, p. 5, https://asiapacific.unfpa.org/sites/default/files/pub-pdf/covid-19_and_vaw_insights_from_big_data_analysis_final.pdf.
40 IPSOS, "Malaysia: Top Issues Faced by Women & Misperceptions of Women Empowerment", 6 April 2018, https://www.ipsos.com/en-my/malaysia-top-issues-faced-women-misperceptions-women-empowerment.
41 Silvia Galdi and Francesca Guizzo, "Media-Induced Sexual Harassment: The Routes from Sexually Objectifying Media to Sexual Harassment", Feminist Forum Review Article, 2020, https://doi.org/10.1007/s11199-020-01196-0; Jaimee Swift and Hannah Gould, "Not an Object: On Sexualization and Exploitation of Women and Girls", UNICEF USA, 11 January 2021, https://www.unicefusa.org/stories/not-object-sexualization-and-exploitation-women-and-girls/30366; "Creating a Healthier Media and Culture", UK Parliament, 23 October 2018, https://publications.parliament.uk/pa/cm201719/cmselect/cmwomeq/701/70107.htm#footnote-117-backlink.
42 Mohd Azizuddin Mohd Sani, Muhammad Zaki Ahmad and Ratnaria Wahid, "Freedom of the Internet in Malaysia", The Social Sciences, 11:7, 1343–1344, 2016, DOI:10.3923/sscience.2016.1343.1349.
43 George (2006) in Sani et al., 2016, p. 1344.

44 Niki Cheong, "Disinformation as a Response to the 'Opposition Playground' in Malaysia", in Sinpeng and Tapsell (eds.) 'From Grassroots Activism to Disinformation', Singapore: ISEAS Publishing, 2021, pp. 66, 69, 79. For Kee and Randhawa ICT's aid to combatting VAW, p. 5.

45 C, H, M, K, G, Haciyakupoglu, pp. 318, 319.

46 C; Ibid, p. 321.

47 A; Ibid, p. 322.

48 H; Ibid, p. 318.

49 O; Ibid, p. 347.

50 F; Ibid, p. 322.

51 C; Ibid, p. 325.

52 S, G, M; Ibid, p. 319.

53 G; Ibid, p. 321.

54 O; Ibid, p. 347. There were also interviewees (B, F, N, K, S, pp. 248–250) who discussed the positive aspects of engaging with Islamic content online – details excluded as they expand beyond this chapter's focus.

55 "Human Rights Commission of Malaysia", item 10, p. 30, 31.

56 World Economic Forum, "Global Gender Gap Report 2021", Insight Report, March 2021, p. 265, https://www3.weforum.org/docs/WEF_GGGR_2021.pdf.

57 "Making Schools a Safer Place: The Malaysian Teen Who Used TikTok to Challenge Abuse", *The Straits Times*, 2 June 2021, https://www.straitstimes.com/asia/se-asia/the-malaysian-schoolgirl-using-tiktok-to-challenge-school-abuse.

58 Ibid.

59 Kee and Randhawa mention ICT use in domestic violence (p. 13).

60 Influenced by the Welsh School's emancipation debate: "alternative reality" can emerge from security's recognition as emancipation, and "true security" can only be materialised by "people and groups who do not deprive others of it". Floyd, 2007, pp. 335, 336.

61 The Malaysian Centre for Constitutionalism and Human Rights (MCCHR), "Cyberharassment in Malaysia: What Do We See Happening?", 31 January 2018, https://mcchr.org/2018/01/31/cyberharassment-in-malaysia-what-do-we-see-happening. UN Women, "Online Violence".

62 "Gender Justice and the Right to Freedom of Opinion and Expression (Malaysia)", A Submission to the Special Rapporteur on the Promotion and Protection of Freedom of Opinion and Expression by KRYSS Network Malaysia with inputs from the Centre for Independent Journalism (CIJ) and Women's Aid Organisation (WAO), 22 June 2021, p. 2, https://www.ohchr.org/sites/default/files/2021-11/KRYSS-Network-Malaysia.pdf

63 UN Women, "Online Violence"; Clarissa Ai Ling and Eric Kerr, "Trolls at the Polls: What Cyberharassment, Online Political Activism, and Baiting Algorithms Can Show Us about the Rise and Fall of Pakatan Harapan (May 2018 – February 2020)", First Monday, 20 May 2020, https://firstmonday.org/ojs/index.php/fm/article/view/10704/9551.

64 UN Women, "Online Violence", p. 29.

65 Ibid, p. 27, 28.

66 "Gender Justice and the Right to Freedom of Opinion and Expression (Malaysia)", p. 2, 4. See also "Voice, Visibility, and A Variety of Viciousness: A Malaysian Study of Women's Lived Realities on Social Media", EMPOWER, 2017, p. 31, 47, https://www.apc.org/sites/default/files/EMPOWER_VVV_FINAL_Web.pdf.

67 "Voice, Visibility, and A Variety of Viciousness: A Malaysian Study of Women's Lived Realities on Social Media", p. 47. *Italics added by the author.*

68 e.g., S, C,T, N, O, H; Haciyakupoglu, p. 330, 331.

69 T, Haciyakupoglu, p. 331.

70 O; Haciyakupoglu, p. 332.

71 H; Haciyakupoglu, pp. 332, 333.

72 H and S; Haciyakupoglu, pp. 333, 334.

73 C; Haciyakupoglu, p. 331.
74 N; Haciyakupoglu, p. 331.
75 Ibid.
76 A, X, P, S, T; Ibid, p. 330.
77 E.g., Sobieraj, 2020.
78 "Voice, Visibility, and A Variety of Viciousness: A Malaysian Study of Women's Lived Realities on Social Media", p. 42.
79 "CENS & the High Commission of Canada Webinar Series on 'Gender, Security and Digital Space: Exploring Risks, Opportunities, and Security Implications' (May 11, 18 & 25)", Panel 2, 10 August 2021, https://www.rsis.edu.sg/rsis-publication/cens/cens-the-high-commission-of-canada-webinar-series-on-gender-security-and-digital-space-exploring-risks-opportunities-and-security-implications/?doing_wp_cron=1628575302.1500000953674316406250#.YRIgUIgzaM.
80 Some securitisation theory discussions influenced my approach: Barry Buzan and Ole Waever, "Macrosecuritisation and Security Constellations: Reconsidering Scale in Securitisation Theory", Review of International Studies, 35: 2, 2009, https://www.jstor.org/stable/20542789; Saleem, 2018; Buzan, Waever and De Wilde in Saleem 2018, p 317; Floyd, 2007.
81 Sani, 2008, p. 92.
82 Sani, 2008, p. 88; Saleem, p. 319. Similarly, Kee and Randhawa discuss how "culture and morality" influence the "frameworks in which women's bodies are regulated to lay claim over national boundaries" (p. 39).
83 Sani, 2008, p. 92.
84 Gender equality advocates may assume different identification and approaches to feminism. See Haciyakupoglu, 2016.
85 Ziba Mir-Hosseini, "Beyond 'Islam' vs. 'Feminism'", IDS Bulletin, 42: 1, 2011, https://doi.org/10.1111/j.1759-5436.2011.00202.x; See discussions in Haciyakupoglu, pp. 144–150.
86 C, F, S, X, Haciyakupoglu, p. 338.
87 E, F, M, O, S, Haciyakupoglu, p. 338.
88 F's quote is from another chapter – included as it provides an explanation of the issue. Haciyakupoglu p. 149.
89 Maznah Mohamad, Cecilia Ng and Tan Beng Hui, "Feminism and the Women's Movement in Malaysia: An Unsung (R)evolution", 2006, New York: Routledge, p. 9.
90 Ibid.
91 Saleem, p. 317.
92 Collins (2005) in Saleem, p. 317, 318.
93 Saleem, p. 325.
94 Lim (2016) in Saleem, p. 325.
95 Saleem, p. 316.
96 Similarly, in a different chapter of the dissertation, an interviewee mentioned how women's organisations had to be informed on Islam to better respond to their "Muslim as well as non-Muslim members and audiences." T Haciyakupoglu, p. 128.
97 This section does not cover the entire legal framework concerning the topic or expand on Syariah (Sharia).
98 Kee and Randhawa, p. 29; Sani, 2008, p. 85, 86; Cheong, p. 73.
99 Sani, 2008, p. 85.
100 Some laws "authorising online surveillance and censorship under the pretext of national security". See "Voice, Visibility, and A Variety of Viciousness: A Malaysian Study of Women's Lived Realities on Social Media", p. 9.
101 Sani et al., 2016. Kee and Randhawa, p. 7, 29.
102 Sani, 2008, p. 88.
103 Cheong, p. 66, 69.

104 Human Rights Watch, "Malaysia: New Government Backslides on Free Speech: Abusive Investigations of Critics Rising", 10 June 2020, https://www.hrw.org/news/2020/06/10/malaysia-new-government-backslides-free-speech.

105 "Gender Justice and the Right to Freedom of Opinion and Expression (Malaysia)", p. 2.

106 "Malaysia", UN Women, and Women Count, https://data.unwomen.org/country/malaysia.

107 UN Women, "Online Violence", p. 40.

108 "The Status of Women's Human Rights", p. 124, 125, 311.

109 "The Status of Women's Human Rights", p. 376, 377

110 Ministry of Communications and Multimedia Malaysia, "Bernama: 21 Jan 2021: Need for ASEAN to Mitigate Cybercrime in Advancing Digital Economy – Muhyiddin", 21 January 2021, https://www.kkmm.gov.my/en/public/news/18470-bernama-21-jan-2021-need-for-asean-to-mitigate-cybercrime-in-advancing-digital-economy-muhyiddin; Majidah Hashim, "COMMENT | Thou Shall not Spread Hate Speech?", Malaysiakini, 22 January 2021, https://www.malaysiakini.com/columns/559961.

111 Hashim, 22 January 2021.

112 "Human Rights Commission of Malaysia", item 7.7, p. 18.

113 F; Haciyakupoglu, pp. 330, 331.

114 H and K; Haciyakupoglu, p. 326.

115 H; Haciyakupoglu, p. 326.

116 C, see above.

117 H; Haciyakupoglu, p. 327.

118 G; Haciyakupoglu, p. 321.

119 X, A, P, T, S; Haciyakupoglu, pp. 329, 330.

120 P, O, M, C, F, S, B; Haciyakupoglu, pp. 339–341.

121 O, C, T, A; Haciyakupoglu, p. 347.

122 N; Haciyakupoglu, p. 349.

123 Various experts highlighted the variance in approaches to online violence and harassment: Nina Jankowicz et al., January 2021, p. 2, 38; Di Meco and Brechenmacher, 30 November 2020; Zeiter et al., 2019, p. 16, 17.

124 Similarly, Kee and Randhawa called for "deeper knowledge" on "how today's context of communication, spaces and technology affect how women experience forms of violence". (p. 37)

125 Similarly, the MCCHR called for laws specifically targeting different types of cyber harassment and UN Women invited states to "Monitor and evaluate ICT VAWG policies, laws and strategies […] their implementation, and initiate modification/reform to accelerate progress where required". MCCHR, 31 Jan 2018; UN Women, "Online Violence", p. 60.

126 "Telenisa", SIS Web page, https://sistersinislam.org/telenisa/.

127 "Telenisa Statistics and Findings", SIS Web page, https://sistersinislam.org/telenisa/.

128 UN Women, "Online Violence", p. 6, 59.

129 Callamard, 2015.

130 Buzan and Waever's (2009) idea of "constellations" can be take into account.

131 UN Women, "Online Violence", p. 58.

132 IPSOS, "International Women's Day 2019", 7 March 2019, https://www.ipsos.com/en-my/international-womens-day-2019.

133 On institutional change: Sobieraj (p. 139, 140) argues that digital attacks are "rooted in social structures and institutional arrangements" and solution demands a change in "institutions and structures that generate them".

134 UN Women, "Online Violence", p. 38.

135 "The Status of Women's Human Rights", Chapter 9, talks about the need to increase women's participation in public and political life.

6

EMOTIONS IN MOTION

Emotions, Viral Justice, and Practices of Security

Yasmine Wong

Throughout history, there is a chronic misrepresentation of women's emotions. Women are often considered "too emotional", resulting in the downplaying of their ability to think and act as rational agents.[1] Women have been excluded from participating in the public sphere on this very basis – that they are "too soft"[2] and thus, not as emotionally suited as men to serve in politics and other positions of power.[3]

Within the wider topic of emotions in political life, there is a tendency to consider emotions with suspicion – to view it as a manifestation of irrationality, an unsophisticated replacement for actual, rational debate.[4] In recent years, however, the idea that emotions are political has become more commonplace.

An arena where women's emotions and the greater emotionality of politics in contemporary times intersect is the #MeToo movement, described to have "brought women's unfiltered pain and anger out into the open".[5] In 2006, survivor and activist Tarana Burke founded "me too" – a movement to bring resources, support, and pathways to healing for survivors of sexual violence.[6] In 2017, the hashtag #MeToo went viral as women shared their experiences with abuse under that hashtag and "came forward to accuse powerful men of harassment and misconduct".[7] This movement shed light on the problem of sexual harassment and assault that plagues societies around the globe, as countries started generating spinoffs specific to local contexts.

In Singapore, conversations on sexual violence and harassment have gained salience more recently, especially in digital spaces. #MeToo has emboldened victims to seek help, with a 79 percent surge in cases reported by Singapore's Sexual Assault Care Centre (SACC).[8]

The case that became iconic of Singapore's #MeToo movement began with a series of stories on Instagram published by Monica Baey, which garnered national outrage at the leniency of her perpetrator's punishment. In 2019, Baey released

DOI: 10.4324/9781003261605-8

stories on her Instagram on the mishandling of a sexual harassment case where her perpetrator was caught filming her in the shower at a campus hostel at the National University of Singapore (NUS). She revealed her perpetrator's personal information and highlighted what she considered to be inadequate punishment meted out to him.[9] Some called the case a "trial by social media", as netizens harassed and shamed the perpetrator online and severely criticised the Singapore Police Force (SPF) and legal system.[10]

This phenomenon has many names. Some refer to it as "digital vigilantism" – a process where citizens, motivated by collective offence, coordinate retaliation on digital platforms and mobile devices.[11] Others call it "viral justice", described by Wood et al. as the result of the virality of a victim's online justice-seeking post.[12] Some situate it within broader online phenomena like "call-out culture"[13] which has been dubbed a "renaissance of public shaming".[14] Viral justice has also been likened to a modern-day witch trial,[15] thematically evoking the idea that emotions, in this case anger, contributes to the same irrationality characteristic of the witch trials, playing out as the upheaval of due process and the subjectification of justice.

Yet, beyond the debate surrounding the breach of integrity of traditional justice systems and speculation on the authenticity of these personal stories of sexual violence, attention should be granted to the underlying politics of emotion that undergird informal processes of justice online. This will be analysed through the lens of vernacular justice and the interpretation of viral justice as security acts in this chapter.

This chapter is inspired by the power of personal narratives at the centre of viral justice, which often feature unfiltered expressions of emotions like "rage, pain, and solidarity",[16] and the idea that emotions capture "conscious thoughts, subjective experiences and normative judgments belonging to the individual".[17] This speaks to the feminist belief that the "personal is political"[18] and the wider recognition that personal narratives on security, and on everyday life, are "never innocent or obvious but always intensely political"[19].[20] As such, the chapter draws attention away from state-centric and militarist notions of (in)security, focusing instead on everyday experiences and expressions of (in)security, specifically vis-à-vis sexual crimes, and the measures taken by individuals and groups to alleviate insecurities.

More specifically, I suggest that an analysis of viral justice through the lens of vernacular security illuminates the power of emotions. This challenges gendered notions of power, of victimhood and agency, and of security as a state-centric process dominated by elite discourse. I first introduce viral justice as the performance – the doing – of vernacular security. Secondly, I highlight the pertinence of emotional narratives in the age of social media. Thirdly, I explore the role of emotions in the formation of agents of security and how this problematises gendered notions of agency and victimhood. Then, I look at emotions and collective security acts. Lastly, I consider the implications of the understanding of emotions as power on existing institutions, and discuss the agenda for change.

Viral justice through the lens of security

In his study of Cameroon, Rogers T. E. Orock situates mob justice within a framework of vernacular security, as a "vernacular technique of securitisation", which emphasises participants' desire for and performance of security amidst perceptions of state and institutional failure to carry out its duty to secure and protect against everyday threats.[21]

Similarly, I will examine viral justice in this chapter as a form of vernacular security, where the performance of security in relation to crimes of a sexual nature is constructed in everyday terms, where citizens are not passive subjects but can and do mould (in)security politics in their daily lives.[22] As in Orock's study, the concept of security here references a state of "being 'safe' either for an individual or collective", where security acts can be defined as practices "directed at procuring and/or maintaining such feelings of 'safety'".[23]

Viral justice involves "an online post 'going viral' and quickly being viewed, shared and re-mediated by large numbers of social media users" and functions "not only to name and shame perpetrators, but also to document evidence".[24] It is a mode of social regulation aimed at perpetrators and the insecurity instigated by their crimes[25] – which is perceived to be inadequately addressed by existing structures of justice and security, and the addressing of which not only increases recognition of the harm, but also aims at restoring a sense of safety.

Vernacular security emphasises that the "vulnerable and insecure" are not pre-determined categories, but individuals and groups who perceive and react to violence in ways that may differ from dominant, state-centric security narratives.[26] This is not to say the source of insecurity – in this case, sexual assault and harassment – does not occur beyond the collective definition of the threat. The problem has always been "out there" but has recently gained significant traction (especially in Singapore) through contestations of state-defined notions of what is and what is not a threat by a collective. This demands a closer inspection of the negotiation of (in)security.

This chapter also draws from the relationship between experiences of victimisation and perceptions of insecurity, where studies often point towards gendered differences. Women generally feel more unsafe and insecure than men, with the fear of sexual violence exceeding the fear of any other crime for women.[27] These insecurities are mirrored in online spaces, where women are disproportionately subjected to various forms of online abuse,[28] particularly online abuse of a sexual nature.[29] In the era of #MeToo, the revelation of widespread violence against women is symptomatic both of patriarchal systems and norms that facilitate crimes against women, and the lack of adequate security responses to women's insecurities.[30]

Digital spaces; spaces of contestation – describing an emotional turn

Social media provides fertile ground to explore how the status of emotions has been elevated as a commodity and a conduit for power. As such, it is useful to begin with an understanding of how social media has contributed, firstly, to the expansion of the space in which political deliberation can take place,[31] and secondly, to the elevation of emotions in practices of vernacular security.

Habermas' (1989) concept of the "public sphere" is defined as a delineated space for citizens to gather and debate affairs on common grounds. It is governed by norms of interaction that facilitate purposeful debate, inevitably reproducing unhelpful binary distinctions – such as assumptions of the public sphere as "rational, impartial, dispassionate and objective".[32] This suggests that "emotionality, partiality, passion, and subjectivity" are unsuited to the public sphere.[33] The conceptualisation of the public sphere as such reveals that it is undergirded by masculine notions of the political and what is worth discussing in public spaces.

Historically, issues such as sexual violence against girls and women are "veiled as 'private' matter[s] without political or public significance".[34] A recent guide on gender norms and women in politics reveals the stubbornness of such beliefs.[35] It argues that although women are increasingly participating in the political sphere, gender norms dictating politics as masculine still function as stumbling blocks.[36]

The persistence of gendered norms can also be seen in the masculine-privileging spaces of mainstream and established media. For example, in Singapore, mainstream headlines in the wake of an exposé[37] on local Indie bookstore BooksActually and the women who had come forward with experiences of inappropriate workplace behaviour by founder Kenny Leck featured how Leck's reputation had been affected and the consequences suffered by the bookstore. This directed attention away from the survivors and their narratives on the issue.

However, as we see in instances of viral justice, women have fashioned "counterpublic online spaces"[38] by using technology to secure "informal justice" through the voicing of personal narratives.[39] Women have used social media to produce and circulate "alternative modes of understanding and responding to gender-based violence".[40] In this sense, new communication technologies, as instruments of social meaning, provide fresh sources of power.[41] Social media provides a means to problematise traditional political reasoning, which privileges rationality, by providing a platform for the sharing of personal experiences. The consequence of this – the production of a new rationality where claims to truth are judged by the speed of communication and emotions engendered.[42] On social media, emotional self-disclosure cultivates genuineness, producing emotional authenticity (regardless of its factual authenticity) which has become a commodity of power.[43] As social media elevates the importance of personal narratives, emotional expressions render the speaker an aura of authenticity and credibility, granting them power to influence public discourse. Social media thus shines a light on and reconfigures the power of emotions as well as the mediation of power relations through emotions. However, it is important to note that the publicity that social media offers may invite victim shaming, as it plays host to debates that feature sexist narratives. For example, in Singapore, when news broke of a 15-year-old girl raped after a game of Truth or Dare, online chatter featured comments about the girl's intoxication and her choice of company, locating at least partial blame on her.[44] This reveals a more nuanced picture of social media as a counterpublic – that it not spared from the gendered norms that exist in traditional spaces.

Emotional subjects; political subjects; subjects of security

After exploring emotions as a commodity on social media, this section argues that recognising the power of emotions in viral justice calls for the problematising of gendered notions of victimhood and political power. This paves the way for new understandings of the securitising agent as emotionally empowered.

Historically, political theory has understood the subject of politics as a rational subject.[45] Postcolonial thought shed light on this "hierarchy between subjects", where "thought and reason are identified with the masculine [] Western subject" and "emotions are associated with the feminine [] racial other".[46] As Sara Ahmed argues, "the projection of emotion onto the bodies of others not only works to exclude others from the realms of thought and rationality, but also works to conceal the emotional and embodied aspects of thought and reason".[47] This ties in with sexist narratives and norms working to discredit women by painting them as emotional and thus unfit for the rationalities of political office. For example, United States Representative Alexandria Ocasio-Cortez was called out by a female journalist for being emotional, with the journalist arguing that her "frequent crying" merely reinforces stereotypes that women are too emotional for politics[48] – a statement that is both informed by and essentialises the idea that women's emotions are necessarily irrational and unpolitical.

This tradition is carried onto security where notions of femininity share a somewhat complex and toxic affinity with victimhood. Security scholars have identified that individuals "classified socially in relation to their race, gender and class" are presented as "objects to be protected or threats to be prevented".[49] Security narratives often capitalise on the feminised victim – who are inherently dependent, vulnerable, and fall short of being full and equal citizens with the capacity to defend themselves.[50] Furthermore, the focus on rape and sexual violence as a global threat has inadvertently prompted concerns that this increase in awareness has perpetuated the image of sexual violence on female-marked bodies, essentialising women as victims and of victims as women.[51]

To frustrate matters, gendered notions of victimhood, and indeed, the labelling of women who share personal experiences of sexual violence and abuse online as "wom[e]n who complain",[52] still hinder the emancipation of women from essentialising categories and serves to curtail emotional expression and participation in online spaces. Victims of sexual harassment and assault and their emotional expressions are often scrutinised, where a "good" rape victim is synonymous to a fallen woman – one who "cannot display their rage or joy or sexuality".[53] As such, the absence of an appearance of suffering in some cases may even lead to the discrediting of the victim, despite emotions being perceived as juridically irrelevant since it is not objective nor factual.[54] For example, despite garnering popular support, some netizens criticised Monica Baey for being "spiteful and vicious" because she chose to speak out against her perpetrator, with some accusing her of acting out of jealousy towards her perpetrator's relationship with his girlfriend.[55]

The policing of emotions – expecting a rational, unemotional political subject, on the one hand, and the strict definition of acceptable emotions for victims, on the other hand – is counterproductive to understandings of security. The sharing of personal narratives by survivors can and should be seen as the description of experiences of insecurity with personal lexicons, cultural contexts, and categories of understanding[56] – an exercise in the defining of a threat seen in vernacular security acts. Although women still face challenges when they share personal stories of abuse online, emotions can serve as the apparatus through which security is performed, challenging masculine, state-centric notions of security, and of who has the power to securitise.

At this point, it is important to clarify that although this chapter has focused on female victims, the feminist task of problematising the relegation of emotions to the private and the feminine is also important in the protection against sexual violence dealt to non-female bodies.

Speaking on male victims of sexual assault and harassment, Anisha Joseph, head of the SACC at the Association of Women for Action and Research (AWARE), mentions that while more male victims are coming forward with their experiences, societal norms still mean that "men and boys are not allowed to express their emotions or are ridiculed for doing so",[57] making it difficult for them to seek help. The association of victims with feminine traits, and of emotional expression as feminine, obscures the recognition of male victims and blocks access to practices of security.

As such, the assertion that emotions are political and are vital to processes of security not only challenges the idea of women as victims, but also of victims as women. By refuting the idea that emotions are inherently "feminine", this allows for a more inclusive movement that incorporates male and non-binary victims. Thus, a feminist examination of sexual violence and (in)security should actively avoid the perpetuation of gendered binaries – of female victims and male perpetrators, and of emotions as feminine and of security as masculine – towards the examination of how gendered notions are reproduced and how this relates to the perpetuation of sexual violence.

Sympathetic collectives; collective practices of security

Complementing the examination of the emotional subject as a securitising agent, this section looks at group practices of security and the function of emotions in the formation of collectives.

When survivors share personal narratives online, they can seek "personal validation, support and recognition from a sympathetic collective, as well as demand [] perpetrator accountability".[58] This section will examine the interaction between the securitising agent and the "sympathetic collective", and how emotions facilitate shared understandings of threats and security acts.

This warrants, firstly, the discussion of viral justice as sites of collective emotional resonance,[59] where emotions are diffused through comments, likes, and the sharing of the personal narratives that resonate with the larger emotional context.[60] This parallels the sociological view of securitisation, where the success of a securitising act is contingent on whether the securitising act resonates with the audience[61] – a process that can be mediated by emotions as a heuristic device and a commodity that grants the securitising agent legitimacy.

When Elicia Yeo realised that her photos had been shared in a Singapore-based Telegram group chat called SG Nasi Lemak where men would circulate obscene photos and videos, and screenshots of girls and make lewd comments on them, she took to Twitter where she saw other young women complaining about the group.[62] Through the airing of their frustrations, sharing of their personal experiences, and through the resonance of emotions, a collective and a collective understanding of (in)security was formed.

Another example is Karmen Siew who took to social media (against the protection of anonymity afforded by the law) to express her anger and disappointment at her offender's lenient sentence after she was molested on public transport.[63] As a result, she received a plethora of messages from others who had similar experiences.[64] This draws attention to the reality that women disproportionately face harassment in public spaces that render them insecure – a reality often overlooked due to underreporting and dominant perceptions of safety in Singapore.[65]

These examples relate to Sara Ahmed's argument about the importance of consciousness raising groups and the coming together of shared experiences. This is because women's testimonies and emotional expressions of pain and are vital "not only to the formation of feminist subjects", but also as a way of "reading pain as structural rather than incidental violence".[66] The compilation of personal experiences and emotional responses to sexual violence online allows the consolidation and definition of a threat, and the understanding of said threat as a systemic problem rather than an individual one. This provides a blueprint to challenge current structures of power.

This leads to, secondly, the idea that the formation of a collective itself can be examined as a security act. The decision to share personal stories of sexual assault and harassment online are at times prompted by the desire to warn others and ensure the safety of the collective. Monica Baey's rationale behind naming her perpetrator was that she never wants him to hurt another person the way she was hurt. Another young woman stated that she posted identifying details of the man who shared her Instagram photos on the SG Nasi Lemak chat to serve as a "warning" to other girls.[67] More recently, the High Court granted an application to lift a gag order on the identity of a Singaporean undergraduate who had filmed voyeuristic videos of almost a dozen female friends after his victims consented to the risk of being identified when his name is published, with some stating the desire to help other victims and warn others.[68] In the United States, a spreadsheet compiling the names and alleged misconduct of men who work in media and publishing was circulated amongst industry professionals, as an attempt by women in the industry to assert themselves "informally, privately", and warn their peers of predatory behaviour.[69] These actions echo what women have been doing for years[70] – from going to the bathroom together to pretending to be a friend to help other women escape precarious situations,[71] women have informally provided safety for one another.

New understandings of security within and beyond existing frameworks

What we have explored is the role of emotions in vernacular security acts. The question remains – what are the implications of viral justice on existing practices of security?

Despite criticism that she disrupted the sanctity of legal processes, Monica Baey's decision to share her experience on social media sparked a nation-wide reckoning with the way cases of sexual harassment and assault are dealt with, especially when technological affordances have disproportionately created new insecurities for women. In response to the backlash, the NUS promised to establish a victim support unit and enhance campus security.[72] A move to review the penalties for three types of sexual offences in Singapore was also prompted by widespread anger and debate over the sentencing of multiple crimes of a sexual nature within universities.[73]

Baey's personal narrative disrupted "official" accounts of justice and security.[74] Her engagement with viral justice at the individual level and the collective response that trailed in the wake of her social media posts reveal the defining and communication of an insecurity, the provision of an informal solution to tackle insecurities unaddressed by existing institutions, and the institutional changes that are prompted through collective action. These expand understandings of security beyond solutions defined and provided by the state and its institutions, and identified areas where institutions have perpetuated the insecurities of survivors. The sharing of her personal experience led to other survivors contacting Baey privately, which helped her identify that survivors tend to avoid official recourse as they felt invalidated by authorities.[75]

The capacity of social media to "represent emotions and provoke strong emotional reactions from other users" challenges the conventional state-centrism of politics and security by redetermining who has the power to participate, what discourses of security have political power, how security is constructed, and by shifting the power dynamics between states and their publics.[76]

Yet, the ability of informal vernacular security acts to challenge state-sponsored definitions and provisions of security has to be weighed against the reality that these insecurities will persist if security elites and institutions do not address them. As Baey argues, the work to lower incidences of sexual assaults and change perceptions "needs to be rooted in governments, institutions and policies", despite the importance of individual efforts.[77]

Still, with the ubiquity of social media in contemporary times, high levels of emotional contagion seem inevitable – and with it, a shared desire for security emerging from shared injustices. What then, can we learn from this?

Implications

Emotions and public morality

To begin more generally, emotions can serve as a gauge for public morality. While emotions are not and should not be the barometer of truth,[78] they can tell us something about justice. The "emotionalisation" of institutions, including legal institutions, is not novel. Anger, disgust, shame, and other negative emotions are seen as "valuable barometers of social morality" and do feature in legal processes.[79] For example, legal processes have the task of "establishing [] justice and fairness that prevent additional arousal of emotions of anger and feelings of revenge" which will serve to enhance acceptance of the judgement by the offender, victim, and the general public.[80]

It is also interesting to note the changing nature of the *public* in conceptualisations of public morality in the age of social media. The internationalisation of "domestic" movements has led to concern over the import of "foreign" concepts of social justice – this includes movements like Black Lives Matter and #MeToo. The former had led to conversations on racism in Singapore, prompting debate when students donning black masks and painted faces drew comparisons to "blackface" – a label considered by some to be borrowed from the West.[81] Such is the globalised nature of the public and public spaces of deliberation in the era of social media. Christine Sylvester's[82] statement on war and emotions can be applied more generally here – that not everyone might have personally experienced violence, but everyone can be touched by it in this era of globalisation. Indeed, injustices that inspire "domestic" movements in one part of the world may find resonance in publics elsewhere, inspiring societies to reckon with their own demons.[83]

It will help to understand that public morality, and indeed, more fundamentally, *publics*, are not stagnant concepts. Improving institutions to incorporate the security of groups that are currently overlooked ensures that society keeps up with public morality – of which institutions should reflect. After all, a "much-needed refresh" of the Singapore's Penal Code was passed only recently in 2019, and with it, the criminalisation of marital rape.[84] (And a series of conversations exacerbated by the glocalisation[85] of the Black Lives Matter movement in Singapore has encouraged authorities to acknowledge the existence of racism in Singapore.)

Emotions, vernacular security, and the agenda for reform

Next, I argue that understanding security through the vernacular, and through the emotions of the public, can help chart institutional reform along more equal lines.

The power of personal narratives and emotions mean the pluralisation of practices of security. This has raised concerns that the legitimacy and monopoly that states and institutional authorities have over violence and security are being threatened by individuals engaging in viral justice.[86] However, vernacular security acts need not be seen as existing in tension with national security and sovereignty. Many cases of viral justice reflect the desire to contest the failure of traditional institutions of justice and security in carrying out their responsibilities and advocate for improvement, rather than the desire for the pluralisation of sovereign control over modes of justice.[87]

Legal institutions have often come under fire for the unjust treatment of women who are survivors of sexual violence for reasons that range from victim blaming and shaming to the leniency granted on grounds of the perpetrator's redeeming qualities. In the infamous Brock Turner trial where Turner was accused of sexually assaulting an intoxicated woman, a lenient sentence was granted on the grounds of Turner being a "first-time offender, promising student and swimming champion".[88] Similarly, in Singapore, Terence Siow, Karmen Siew's perpetrator, had his offences described by the district judge as "minor intrusions", and his academic results were used as evidence that he had "potential to excel in life".[89] Monica Baey's perpetrator was given a conditional warning instead of a sentence due to a "high likelihood of rehabilitation" and because he "was remorseful".[90]

In addition, these statements often overlook the damages faced by survivors of sexual abuse and of the fallout that trails the sharing of their personal narratives online. In an example from the United States, a woman shared her account of being fired from her job because her employer came across defamatory content posted by her harasser online and wanted to avoid controversy.[91] The damage to women's reputations and economic lives may lead to self-censorship, deterring other survivors from seeking justice. This remains under-investigated in Singapore, although it is reported that women who experience sexual harassment at work are often discouraged from reporting due to the fear of retaliation, suggesting that women still face consequences when they seek recourse.[92]

The perceived legitimacy of states is highly dependent on the state's ability to fulfil core functions of security.[93] As such, it is important for Singapore – and of course, for all countries – to acknowledge and address how institutions have failed to ensure the security concerns of women and other marginalised communities. Along these lines, it is essential to scrutinise our institutions and understand their role in the (in)security of women. As products of human agency and norms, institutions often ossify existing structures of inequality and insecurities in place. Foucault makes a similar argument against institutional universals, arguing that any set of social arrangements, as institutions are, rests on a "particular, arbitrary, conceptual structure" that is a "regime of power".[94] The view of institutions as social constructs challenges the idea that they are fixed entities and provides grounds for security to be understood from other, marginalised points of view, such as that of women who seek justice through means other than that of established institutions.

In Singapore, recent concern over rampant sexual misconduct in universities sparked by Monica Baey's case has led to compulsory courses on sexual consent.[95] Such institutional support is key to changing norms and attitudes in how we view the security of women and an example of institutional and cultural shifts along more equal lines. The updating of laws to account for the insecurities of women also sends a wider message to the public as they establish the boundaries of acceptable behaviour. The reassessment of the penalties for sexual offences in Singapore, for example, reflects recognition of the severity of these crimes and its impact on victims and proves that the consultation of survivors is important when designing policy solutions that address unique insecurities. To highlight this point, an anti-molestation campaign by the SPF left a bad taste because the posters – which featured a hand reaching to touch another person (insinuating molest) with a tag on the wrist that says "2 years' imprisonment. It's not worth it" – neglected to centre the harm inflicted on the survivors.[96]

Admittedly, there has been scepticism locally over viral justice. The increased prevalence of doxxing and shaming that often follow in the wake of individuals sharing their experiences with injustices online led to the inclusion of "doxxing" as an offence in Singapore.[97] This is an important step in demarcating the limits viral justice. To clarify, the demand for acknowledgement of emotions in institutional processes and the obscuring of the binary between "emotional" and "rational" does not mean the relinquishing of institutional sovereignty. As such, it is important for a balance to be struck between acknowledgement and curtailment when it comes to informal justice.

Social media and the cure

Before concluding, I will grant attention to the complicated relationship between social media and equality. While publicity allows women to take control of the narrative, visibility also poses its own set of insecurities. The domination of the male gaze means that the visibility of women and individuals of marginalised identities online often invites objectification. As such, despite the optimism that social media provides a platform for vernacular security acts, it is important to acknowledge that the use of social media ultimately comes at a risk for women, especially survivors of sexual harassment and assault whose sharing of personal experiences may render them vulnerable to re-victimisation.

Next, the ability of social media as a counterpublic to circulate alternative modes of speech, understandings, and actions should not be overstated. Such content can only "circulate up to a point, at which it is certain to meet resistance".[98]

As such, the factoring of emotions must resist limited focus on women as victims or as providers of their own security – the former ignoring women's active efforts to seek security and the latter placing overt emphasis on individual responsibility. Despite social media serving as a counterpublic (not without its own set of limitations) and its role in security acts, viral justice and the affordances of social media should be seen as diagnostic of, rather than a solution to,[99] violence against women. This is in line with Fraser's argument that systemic inequality must be eliminated to achieve participatory parity.[100]

One fundamental question to ask is – why are women and other marginalised groups turning to social media to exact informal justice in the first place? One can say that social media affords the ability for individuals to broadcast and produce their own content, unlike that of traditional institutions and public spaces. This is evidenced by the physical town hall session at the NUS following the Monica Baey case, where several NUS students were angered as they felt that the session did not allow them to air their concerns and participate in policymaking regarding sexual misconduct,[101] pointing to the fact that official spaces for dialogue still fall short of validating individual experiences. This calls attention to the need for spaces for legitimate dialogue.

Lastly, while emotions are quintessential in politics, they may also serve as pressure points to exploit, thus a balance should be sought. In the age of social media, moral sentiments and emotions have transcended interactions between individuals and the larger social context of communities – and because of this lack of mutuality and duration, they become volatile.[102] Research has explored the exploitation of emotions – particularly, the grievances and feelings of nostalgia of the populace – by populist movements and the far right.[103] In addition, anger has been suggested to promote belief in misinformation motivated by political alignment, and anxiety has been linked to belief in politically divisive fake news.[104] Further inquiries would have to address where we draw the line – what we consider to be appropriate and acceptable when it comes to emotions in politics. While there is no easy way to circumvent this, the ability to engage in emotional regulation, and the capacity to empathise with the emotions of others aids in the detection of fake news. As such, a consideration of emotions and emotional intelligence can be incorporated into the development of digital literacy programmes.[105]

Conclusion

In an Instagram post seeking closure, Monica Baey addressed the NUS's acknowledgement of the inadequacies in the way they deal with sexual misconduct cases and reflected on her actions amidst discussions of deservedness and justice. Like many others, her expressions of anger and hurt paved the way for institutional reflection and change, and the promise of an understanding of security that better encapsulates the (in)securities of women, and of victims of sexual misconduct regardless of gender.

Since #MeToo took the world by storm, survivors around the world continue to share their experiences with sexual violence via social media, prompting reflection at various levels and inspiring new waves of activism. In Australia, the movement finally made its way into Australian politics, sparking protests and inquiries into parliamentary work culture,[106] while in countries like China, the movement still grapples with heavy censorship.[107] In Singapore, change has only just begun, with the government announcing the presentation of a White Paper in 2022 on issues concerning women, including the protection of women from harassment and insecurities online.[108]

Even when progress is stymied amidst existing power structures that shield abusers from processes of justice, the courage to tell stories that depict personal experiences with sexual violence continue to inspire a more nuanced understanding of security and justice that centres the experiences of survivors. Bernice Fisher states that "the voices that make us the most uncomfortable and the feelings that accompany them constitute a built-in critique of our ideals". Perhaps, what is needed is a deeper understanding and balance when it comes to the role of emotions in politics, or emotions *as* politics.

Acknowledgement

A big thank you to Dr Tamara Nair and Benjamin Ang. Their comments helped get the chapter to where it stands today.

Notes

1 Dean Burnett, "Calm Down, Dear: The Dark Side of 'Emotional Intelligence'", *The Guardian,* 21 April 2015, https://www.theguardian.com/science/brain-flapping/2015/apr/21/calm-down-dear-emotional-intelligence-psychology
2 Divya Rajagopal, "It's Assumed Women Are too Soft, Emotional or Hysterical for Leadership Roles: Julia Gillard", *The Economic Times,* 6 February 2019, https://economictimes.indiatimes.com/magazines/panache/its-assumed-women-are-too-soft-emotional-or-hysterical-for-leadership-roles-julia-gillard/articleshow/67861237.cms?utm_source=contentofinterest&utm_medium=text&utm_campaign=cppst
3 Susan Milligan, "Women Candidates Still Tagged as Too 'Emotional' to Hold Office", *U.S. News,* 16 April 2019, https://www.usnews.com/news/politics/articles/2019-04-16/women-candidates-still-tagged-as-too-emotional-to-hold-office; Victoria L. Brescoll, "Leading with Their Hearts? How Gender Stereotypes of Emotion Lead to Biased Evaluations of Female Leaders", *The Leadership Quarterly,* 27(2016), 415–428.
4 Lauren K. Hall, "Review: Impassioned Politics: New Research on the Role of Emotions in Political Life", *Politics and the Life Sciences,* 28(2) (2009), 84.
5 Olivia Adams, "The #MeToo Movement Needs to Stay Angry, if We Are to Deliver Real Change for Women", *Marie Claire,* 29 January 2020, https://www.marieclaire.co.uk/opinion/me-too-movement-women-violence-harvey-weinstein-683309
6 "History and Inception", *metoomvmt.org,* n.d., https://metoomvmt.org/get-to-know-us/history-inception/
7 "#MeToo Is at a Crossroads in America. Around the World, it's just Beginning", *The Washington Post,* 8 May 2020, https://www.washingtonpost.com/opinions/2020/05/08/metoo-around-the-world/

8 Christine Amour-Levar, "#MeToo Movement in Asia: Is Singapore Feeling the Weinstein Effect?", *Forbes,* 17 December 2017, https://www.forbes.com/sites/christineamourlevar/2017/12/17/sexual-harassment-in-the-workplace-is-singapore-feeling-the-weinstein-effect/?sh=783229e6160d

9 Matthias Ang, "NUS Student who Revealed Male Perpetrator's Details Could Be Breaking Proposed Doxing Laws", *Mothership,* 2 April 2019, https://mothership.sg/2019/04/nus-student-doxxing-peeping-tom-harassment/

10 Yasmine Wong, "Victim and the Cyber Vigilante: An Additional Perspective on Cyber Vigilantism", in Majeed Khader, Loo Seng Neo and Whistine Xiau Ting Chai (Eds.), *Introduction to Cyber Forensic Psychology: Understanding the Mind of the Cyber Deviant Perpetrators,* Singapore: World Scientific (2021).

11 Daniel Trottier, "Digital Vigilantism as Weaponisation of Visibility", *Philosophy and Technology,* 30(1) (2017), 55–72.

12 Mark Wood, Evelyn Rose, and Chrissy Thompson, "Viral Justice? Online Justice-Seeking, Intimate Partner Violence and Affective Contagion", *Theoretical Criminology,* 23(3) (2019), 375–393.

13 Ruth M Dunsby and Loene M Howes, "The NEW Adventure of the Digital Vigilante! Facebook Users' Views on Online Naming and Shaming", *Journal of Criminology,* 42(1) (2019); Asam Ahmad, "A Note on Call-Out Culture", *Briarpatch,* 2 March 2015, https://briarpatchmagazine.com/articles/view/a-note-on-call-out-culture

14 Jon Ronson, *So You've Been Publicly Shamed,* London: Pan Macmillan (2015); ibid

15 Robert T Muller, "Online Humiliation and the Shame It Brings", *Psychology Today,* 10 December 2020, https://www.psychologytoday.com/intl/blog/talking-about-trauma/202012/online-humiliation-and-the-shame-it-brings

16 Allison Page and Jacquelyn Arcy, "#MeToo and the Politics of Collective Healing: Emotional Connection as Contestation", *Communication, Culture and Critique,* 13(3) (2020), 333.

17 Linda Åhäll, "Affect as Methodology: Feminism and the Politics of Emotion", *International Political Sociology,* 12(1) (2018), 36–52.

18 Carol Hanisch, "The Personal is Political", *carolhanisch.org,* February 1969, http://www.carolhanisch.org/CHwritings/PIP.html

19 Annick T. R. Wibben, *Feminist Security Studies: A Narrative Approach.* Abingdon: Routledge (2011), 2.

20 Lee Jarvis, "Toward a Vernacular Security Studies: Origins, Interlocutors, Contributions, and Challenges", *International Studies Review,* 21(1) (2019), 109.

21 Rogers T. E. Orock, "Crime, In/Security and Mob Justice: The Micropolitics of Sovereignty in Cameroon", *Social Dynamics,* 40(2) (2014), 408–428.

22 Nick Vaughan-Williams and Daniel Stevens, "Vernacular Theories of Everyday (in)Security: The Disruptive Potential of Non-elite Knowledge", *Security Dialogue,* 47(1) (2016), 53.

23 Ibid, 411.

24 Mark Wood, Evelyn Rose, and Chrissy Thompson, "Viral Justice? Online Justice-Seeking, Intimate Partner Violence and Affective Contagion", *Theoretical Criminology,* 23(3) (2019).

25 Orock, "Crime, In/Security and Mob Justice", 409.

26 Robin Luckham, "Whose Violence, Whose Security? Can Violence Reduction and Security Work for Poor, Excluded and Vulnerable People?", *Peacebuilding* 5(2) (2017), 112; Lee Jarvis, "Toward a Vernacular Security Studies: Origins, Interlocutors, Contributions, and Challenges", *International Studies Review,* 21(1) (2019), 108.

27 Oksan Tandogan and Bige Simsek Ilhan, "Fear of Crime in Public Spaces: From the View of Women Living in Cities", *Procedia Engineering,* 161 (2016), 2013.

28 Azmina Dhrodia, "OPINION: To Stop Online Abuse against Women, We must Reform Digital Spaces", *Thomson Reuters Foundation,* 9 April 2021, https://news.trust.org/item/20210409123542-l58r0

29 Maeve Duggan, "Men, Women Experience and View Online Harassment Differently", *Pew Research Center*, 14 July 2017, https://www.pewresearch.org/fact-tank/2017/07/14/men-women-experience-and-view-online-harassment-differently/

30 Marissa Fortune, "How the MeToo Movement Highlights the Need for Security Sector Reform in the Global North", *ISSAT Blog*, 30 November 2018, https://issat.dcaf.ch/sqi/Share/Blogs/ISSAT-Blog/How-the-MeToo-Movement-Highlights-the-Need-for-Security-Sector-Reform-in-the-Global-North

31 Constance Duncombe, "The Politics of Twitter: Emotions and the Power of Social Media", *International Political Sociology*, 13(4) (2019), 410.

32 Karin Wahl-Jorgensen, "Questioning the Ideal of the Public Sphere: The Emotional Turn", *Social Media + Society*, July-September (2019), 2.

33 Ibid; Nancy Fraser, "Rethinking the Public Sphere: A Contribution to the Critique of Actually Existing Democracy", *Social Text*, 25/26 (1990), 59.

34 Carole Pateman, *The Sexual Contract*, California: Stanford University Press, 1988.

35 Rachel George, Emma Samman, Katie Washington and Alina Ojha, "Gender Norms and Women in Politics: Evaluating Progress and Identifying Challenges on the 25th Anniversary of the Beijing Platform", *ALiGN*, August 2020.

36 Ibid.

37 Charmaine Poh, "Built on Uneasy Compromises: The Young Women behind BooksActually Speak Up", *Rice*, 25 September 2021, https://www.ricemedia.co/books-actually-young-women-speak-up/

38 A "counterpublic" reflects an expansion on Habermas' notion of a single public sphere, denoting "parallel discursive arenas where members of subordinated social groups invent and circulate counter discourses, so as to formulate oppositional interpretations of their identities, interests and needs". See: Nancy Fraser, "Rethinking the Public Sphere: A Contribution to the Critique of Actually Existing Democracy", *Social Text*, No. 25/26 (1990), 56–80.

39 Ibid.

40 Michael Salter, "Justice and Revenge in Online Counter-Publics: Emerging Responses to Sexual Violence in the Age of Social Media", *Crime Media Culture*, 9 (3) (2013), 228.

41 Nicola Henry and Anastasia Powell, "Embodied Harms: Gender, Shame, and Technology-Facilitated Sexual Violence", *Violence Against Women*, 21(6) (2015), 762.

42 Duncombe, "The Politics of Twitter", 421.

43 Karin Wahl-Jorgensen, *Emotions, Media and Politics*, New Jersey: John Wiley & Sons (2019).

44 Shailey Hingorani, "Commentary: She's Practically Asking for It? Do Singaporeans Subscribe to Rape Myths?", *Channel News Asia*, 4 November 2019, https://www.channelnewsasia.com/commentary/singapore-sexism-rape-sexual-assault-upskirt-why-so-many-cases-847181

45 Inder S. Marwah, "What Nature Makes of Her: Kant's Gendered Metaphysics", *Hypatia*, 28(30) (2013), 551–567; Duncombe, 415.

46 Sara Ahmed, *The Cultural Politics of Emotion*, 2nd Edition, New York: Routledge (2015).

47 Ibid, 170.

48 Becca Blaser, "Women and Politics: Too Emotional?", *International Policy Digest*, 22 November 2019, https://intpolicydigest.org/women-and-politics-too-emotional/

49 Mariana Selister Gomes and Renata Rodrigues Marques, "Can Securitization Theory Be Saved from Itself? A Decolonial and Feminist Intervention", *Security Dialogue*, 52(5) (2021), 79.

50 Lauren Wilcox, "Gendering the Cult of the Offensive", in Laura Sjoberg (Ed.), *Gender and International Security: Feminist Perspective*, New York: Routledge (2010), 75.

51 Paula Drumond, Elizabeth Mesok and Marysia Zalewski, "Sexual Violence in the Wrong(ed) Bodies: Moving beyond the Gender Binary in International Relations", *International Affairs*, 96(5) (2020), 1145.

52 Sarah Sobieraj, *Credible Threat: Attacks Against Women Online and the Future of Democracy,* New York: Oxford University Press, (2020), 99.

53 Jane Doe, *The Story of Jane Doe: A Book About Rape,* Canada: Random House Canada (2003), 118.

54 Isabel Ventura, "'They Never Talk about a Victim's Feelings: According to Criminal Law, Feelings Are not Facts'—Portuguese Judicial Narratives about Sex Crimes", *Palgrave Communications,* Palgrave Macmillan, 2(1) (2016), 1–9.

55 "These Online Comments Blaming Peeping Tom Victim Monica Baey and Supporting her Perpetrator Will Make You Throw Up", *Coconuts Singapore,* 28 April 2019, https://coconuts.co/singapore/news/these-online-comments-blaming-peeping-tom-victim-monica-baey-and-supporting-her-perpetrator-will-make-you-throw-up/

56 Lee Jarvis, "Toward a Vernacular Security Studies: Origins, Interlocutors, Contributions, and Challenges", *International Studies Review,* 21(1) (2019), 108; Stuart Croft and Nick Vaughan-Williams, "Fit for Purpose? Fitting Ontological Security Studies 'into' the Discipline of International Relations: Towards a Vernacular Turn?", *Cooperation and Conflict,* 52(1) (2017).

57 Ang Hwee Min, "Male Molest Numbers Continue to Rise; Experts Say some Victims Struggle to Report Cases", *Channel News Asia,* 26 October 2019, https://www.channelnewsasia.com/singapore/male-victims-molest-outrage-modesty-singapore-cases-1313661

58 Mark A. Wood, Evelyn Rose and Chrissy Thompson, "Viral Justice? Online Justice-Seeking, Intimate Partner Violence, and Affective Contagion", *Theoretical Criminology,* 23(3) (2019), 375–393.

59 Duncombe, "The Politics of Twitter", 415.

60 Thierry Balzacq, "The Three Faces of Securitization: Political Agency, Audience and Context", *European Journal of International Relations,* 11(2) (2005), 182.

61 Ibid, 182; Eric Van Rythoven, "Learning to Feel, Learning to Fear? Emotions, Imaginaries, and Limits in the Politics of Securitisation", *Security Dialogue*, 46(5) (2015), 459.

62 Kimberly Lim, "How Young Women Are Using Social Media to Fight Back against Men Behaving Badly", *Today,* 12 October 2019, https://www.todayonline.com/singapore/how-young-women-are-using-social-media-fight-back-against-men-behaving-badly

63 Tanya Ong, "Woman Shares Frustration about Being Molested on NEL Train & Escalator at Serangoon MRT Station, as Molester Got Away", *Mothership,* 14 September 2018, https://mothership.sg/2018/09/molester-serangoon-mrt-got-away/

64 Ibid.

65 Olivia Lin and Mandy How, "S'pore Is Safe but Creeping on Girls & Women in Public Is not Okay, Okay?", *Mothership,* 16 May 2017, https://mothership.sg/2017/05/spore-is-safe-but-creeping-on-girls-women-in-public-is-not-okay/

66 Ahmed, *The Cultural Politics.*

67 Kimberly Lim, "How Young Women Are Using Social Media to Fight Back against Men Behaving Badly", *Today,* 12 October 2019, https://www.todayonline.com/singapore/how-young-women-are-using-social-media-fight-back-against-men-behaving-badly

68 Louisa Tang, "Chief Justice Lifts Gag Order on Identity of Singaporean Undergrad Voyeur from Top UK University", *Today,* 24 September 2021, https://www.todayonline.com/singapore/chief-justice-lifts-gag-order-identity-singaporean-undergrad-voyeur-top-uk-university

69 Alana Massey, "Women Have always Tried to Warn Each Other about Dangerous Men. We Have to", *The Washington Post,* 13 October 2017, https://www.washingtonpost.com/news/posteverything/wp/2017/10/13/women-have-always-tried-to-warn-each-other-about-dangerous-men-we-have-to/

70 Ibid.
71 Tanya Chen, "A Woman Shared a Moment She Helped a Female Stranger Escape a Stalker and the Responses Are Chilling", *Buzzfeed News,* 25 July 2018, https://www.buzzfeednews.com/article/tanyachen/women-sharing-moments-they-felt-threatened-were-stalked
72 Amelia Teng, "NUS Admits It Failed Monica Baey; Will Set Up Victim Support Unit and Improve Campus Security", *The Straits Times,* 26 April 2019, https://www.straitstimes.com/singapore/education/nus-admits-it-failed-monica-baey-will-set-up-victim-support-unit-and-improve
73 Lydia Lam and Jalelah Abu Baker, "Penalties for 3 Sex Crimes to Go Up after Review, Academic Potential Should not Carry Much Weight: Shanmugam", *Channel News Asia,* 5 March 2021, https://www.channelnewsasia.com/singapore/penalties-for-3-sex-crimes-to-go-up-after-review-shanmugam-308926
74 Vaughan-Williams and Stevens, "Vernacular Theories", 40–58.
75 Wong Pei Ting, "Victimised Women Continue to Engage Monica Baey, Who Says Casual Attitude towards Voyeurism Must Stop", *Today,* 28 August 2020, https://www.todayonline.com/singapore/victimised-women-continue-engage-monica-baey-who-says-casual-attitude-towards-voyeurism
76 Duncombe, "The Politics of Twitter", 410.
77 Rei Kurohi, "Both Legislation and Mindset Shifts Needed to Counter Toxic Masculinity: IPS Panel", *The Straits Times,* 3 June 2021, https://www.straitstimes.com/singapore/both-legislation-and-mindset-shifts-needed-to-counter-toxic-masculinity-panel
78 Olivia Goldhill, "2018 Is the Year Women Tried to Reclaim Anger but Failed", *Quartz,* 21 December 2018, https://qz.com/1496726/2018-is-the-year-women-tried-to-reclaim-anger-but-failed/
79 Susanne Karstedt, "Emotions and Criminal Justice", *Theoretical Criminology,* 6(3) (2002), 308.
80 Ibid.
81 Jane Zhang, "Ong Ye Kung: Racial Insensitivities Exist in S'pore, But Our Situation 'Entirely Different' from US", *Mothership,* 6 June 2020, https://mothership.sg/2020/06/racial-microaggressions-ong-ye-kung/
82 Christine Sylvester, "War, Sense, and Security", in Laura Sjoberg (Ed.), *Gender and International Security: Feminist Perspective,* New York: Routledge (2010), 25.
83 Norman Vasu and Yasmine Wong, "BLM Movement: Singapore and Glocalisation", *RSIS Commentary-* CO20141, 13 July 2020, https://www.rsis.edu.sg/rsis-publication/cens/blm-movement-singapore-and-glocalisation/#.YercA_5ByM8
84 Kevin Kwang and Aaron Chong, "Sweeping Law Reforms to Outlaw Marital Rape, Penalise Voyeurism Passed", *Channel News Asia,* 6 May 2019, https://www.channelnewsasia.com/singapore/law-reforms-passed-outlaw-marital-rape-penalise-voyeurism-876526
85 Vasu and Wong, "BLM Movement".
86 Orock, "Crime, In/Security and Mob Justice", 420.
87 Ibid, 408–428.
88 Lynn Neary, "Victim of Brock Turner Sexual Assault Reveals Her Identity", *NPR,* 4 September 2019, https://www.npr.org/2019/09/04/757626939/victim-of-brock-turner-sexual-assault-reveals-her-identity; Elle Hunt, "'20 Minutes of Action': Father Defends Stanford Student Son Convicted of Sexual Assault", *The Guardian,* 6 June 2016, https://www.theguardian.com/us-news/2016/jun/06/father-stanford-university-student-brock-turner-sexual-assault-statement
89 "'Potential to Excel in Life': NUS Undergrad Who Molested Woman Gets Probation for 'Minor Intrusion' Offences", *The Straits Times,* 26 September 2019, https://www.straitstimes.com/singapore/courts-crime/university-student-who-molested-woman-gets-probation-for-minor-intrusion

90 Amelia Teng, "NUS Peeping Tom Given Conditional Warning due to High Likelihood of Rehabilitation: Police", *The Straits Times,* 23 April 2019, https://www.straitstimes.com/singapore/courts-crime/student-in-nus-sexual-misconduct-case-given-conditional-warning-due-to-high

91 Sobieraj, *Credible Threat,* 107.

92 Anna Maria Romero, "Majority of Perpetrators of Sexual Harassment at Work Suffer no Consequences — AWARE", *The Independent,* 11 January 2022, https://theindependent.sg/majority-of-perpetrators-of-sexual-harassment-at-work-suffer-no-consequences-aware/

93 Orock, "Crime, In/Security and Mob Justice", 418.

94 Mark Bevir, "Foucault, Power, and Institutions", *Political Studies,* 47(2) (1999), 352.

95 Meera Navlakha, "Certain Singapore Universities Have Seen an Increase in Sexual Misconduct Cases in Recent Years", *Vice,* 27 August 2019, https://www.vice.com/en/article/8xwgn5/top-singapore-universities-now-have-compulsory-courses-on-sexual-consent

96 Adeline Tan, "Victims Weigh in on Anti-Molestation Posters", *The New Paper,* 25 November 2019, https://tnp.straitstimes.com/news/singapore/victims-weigh-anti-molestation-posters

97 Lianne Chia, "'Doxxing' to Be Criminalised under Amendments to Protection from Harassment Act", *Channel News Asia,* 1 April 2019, https://www.channelnewsasia.com/news/singapore/doxxing-to-be-criminalised-under-amendments-to-protection-from-11400756

98 Michael Warner, "Publics and Counterpublics", *Public Culture,* 14(1) (2002), 87.

99 Emma A. Jane, "Feminist Digilante Responses to a Slut-Shaming on Facebook", *Social Media +Society,* April-June (2017), 1–10.

100 Ioannis Kampourakis, "Nancy Fraser: Subaltern Counterpublics", *Critical Legal Thinking,* 6 November 2016, https://criticallegalthinking.com/2016/11/06/nancy-fraser-subaltern-counterpublics/

101 Meera Navlakha, "Top Singapore Universities Now Have Compulsory Courses on Sexual Consent", *Vice,* 27 August 2019, https://www.vice.com/en/article/8xwgn5/top-singapore-universities-now-have-compulsory-courses-on-sexual-consent

102 Susanne Karstedt, "Emotions and Criminal Justice", *Theoretical Criminology,* 6(3) (2002), 304.

103 Nicolas Demertzis, "Emotions and Populism", in Simon Clarke, Paul Hoggett and Simon Thompson (Eds.), *Emotion, Politics and Society,* New York: Palgrave Macmillan (2006), 114.

104 Brian E. Weeks, "Emotions, Partisanship, and Misperceptions: How Anger and Anxiety Moderate the Effects of Partisan Bias on Susceptibility to Political Misinformation", *Journal of Communication,* 65(4) (2015), 699–719.

105 Tony Anderson and David James Robertson, "Fake News: People with Greater Emotional Intelligence Are Better at Spotting Misinformation", *The Conversation,* 23 March 2021, https://theconversation.com/fake-news-people-with-greater-emotional-intelligence-are-better-at-spotting-misinformation-157265

106 Blair Williams, "In 2021 #MeToo Finally Made It to #Auspol—What Happens Next?", *The Conversation,* 21 December 2021, https://theconversation.com/in-2021-metoo-finally-made-it-to-auspol-what-happens-next-173153

107 Lü Pin, "What Peng Shuai's Story Tells Us about MeToo in China", *The Diplomat,* 5 November 2021, https://thediplomat.com/2021/11/what-peng-shuais-story-tells-us-about-metoo-in-china/

108 Jalelah Abu Baker, "Government to Study Views on Women's Issues, Present 'Concrete Proposals' in White Paper in Early 2022: PM Lee", *Channel News Asia,* 18 September 2021, https://www.channelnewsasia.com/singapore/pm-lee-women-issues-white-paper-early-2022-2186861

PART III

7

NASTY, FAKE AND ONLINE

Distinguishing Gendered Disinformation and Violence Against Women in Politics

Gabrielle Bardall

Where women speak up to lead and to dissent in the public sphere, they regularly face abusive and violent backlash. In Canada, Catherine McKenna (former Minister of the Environment), Sandra Jansen (Alberta MLA) and Rachel Notley (current Premier of Alberta) have been subject to record levels of threats and online harassment, sometimes spilling into offline violence.[1] This may be exacerbated by other elements of their identity, as with the case of Kathleen Wynne (former Premier of Ontario), who became the target of harassment as a woman and openly lesbian politician.[2] In the United States, Kamala Harris (the first racialized and first woman Vice President), Hillary Clinton (former Presidential candidate, Governor of New York and First Lady) and Sarah Palin (former Vice Presidential candidate and Governor of Alaska) have all faced intense sexualized depictions, including so-called "deep fakes" on social media.[3] In the United Kingdom, MP Jess Phillips reported receiving over 600 rape and death threats online in a single night.[4]

Online harassment and violence against political women are not limited to advanced democracies. In Rwanda, fake nude photos were used as a smear tactic against presidential candidate Diane Rwigara.[5] In Ukraine, MP Svitlana Zalishchuk was taunted for years by a sexualized false rumor on Twitter after she gave a speech at the United Nations (UN) about women in conflict.[6] In Argentina, Ofelia Fernandez, the youngest politician to be elected to Buenos Aires city legislature, cited online abuse and harassment as the reason she had to withdraw from social media[7] as did Indian women's rights activist, Kavita Krishnan,[8] British MP Flick Drummond[9] and innumerable others.

Collectively, these cases can be globally identified as examples of **violence against women in politics** (VAWP) or, more accurately, violence against women in public life. Violence and harassment against women in politics is an insidious form of political violence that systematically maintains patriarchal

DOI: 10.4324/9781003261605-10

power in state institutions. It is also a profoundly harmful form of gender-based violence (GBV) in terms of its impacts on women's health. By preventing women from acceding to decision-making positions, it allows patterns of GBV to perpetuate. Because of this, GBV targeted at women leaders in this way has a doubly profound effect: it both harms the individual woman and diminishes or removes her voice from public leadership where she would have been in a position to disrupt broader cycles of violence.

A large-scale uptick in global awareness about these online attacks began to crescendo at the same time as another popular catchphrase, "disinformation", captured public attention. "Disinformation" became a household word following the election of Donald Trump. The origins of the term pre-date the internet by many decades. Yet, it was quickly and inextricably associated with social media in popular perception. The coincident prominence of these two streams of toxic online behavior prompted important work on understanding potential relationships between them and produced a new concept, "gendered disinformation". Today, "gendered disinformation" and "violence against women in politics" are buzzwords in studies of political manipulation and abuse in both advanced and transitional democracies. However, conflating the issues muddies the path to effective remedy.

In this chapter, I argue that the willful and malicious spreading of false information based on harmful gendered stereotypes ("gendered disinforming") is a form of VAWP. Yet, common assumptions and misperceptions about disinformation have clouded the concept. This chapter breaks down the concepts of VAWP and gendered disinformation to identify where they relate to each other and how more precision can contribute to more effective application of the concepts.

Defining online violence against women in politics and gendered disinformation

"Gendered disinformation" should be understood as a subset of online VAWP, which, in turn is just one form of VAW. VAWP is any act, or threat, of physical, sexual or psychological GBV against women that prevents women from exercising and realizing their political rights and a range of human rights.[10] Acts of VAWP range from femicide to marital rape to psychological abuse and family violence. VAWP occurs both online and offline.

Online VAWP may come in various intensities, ranging from general acts of incivility to outright threats.[11] **Incivility** refers to aggressive and disrespectful behaviors, harassment, hate speech and outrageous claims that are spread on social networking sites. Incivility can have a particularly harmful effect on women with multiple marginalized identities.[12] **Information disorder** (including disinformation, misinformation and mal-information) is another uncivil behavior. It may mask overt expressions of harm behind apparently innocuous or humorous content or intentionally spread both true and false damaging information under the guise of impartial reporting. **Online hate** (or online hate speech), on the other hand, is defined as any online expression, encouragement, stirring up or incitement of hatred.[13] Hate speech vilifies, humiliates or incites hatred against a group or a class of persons/based on a protected attribute such as the target's sexual orientation, gender, religion, disability, color or country of origin.[14] One of the most intense forms of online VAWP is overt and direct threats of **physical or sexual harm**. Other forms of criminally harmful behaviors fall into this group as well, such as criminal harassment, defamation and stalking.

As per this chapter's focus on gendered disinformation, it is essential to unpack what "Information Disorder" entails. While defining the components of information disorder helps with conceptual clarity, the chapter puts more emphasis on disinformation because it is the most widely used concept. "Information disorder", the sharing of fake and/or harmful information,[15] is an effective way of weaponizing information and perceptions to achieve political ends. Smear tactics and character assassination have been tricks of dirty politics since time immemorial. Yet, today information disorder has become largely associated with online communications, especially social media. Information disorder includes misinformation, disinformation and mal-information.[16]

The term "disinformation" is frequently incorrectly used as a catch-all for multiple forms of information disorder. **Disinformation** is false information that is shared knowingly to cause harm. It is an intentional deception with malicious intent. **Misinformation** is false information that one spreads without knowing it is false or because one believes it to be true. It is shared with no intent to cause harm although its effects can be very harmful indeed. **Mal-information** is when genuine information is shared to cause harm, often by moving information designed to stay private into the public sphere.[17] Gendered misinformation can be an outcome of gendered disinformation and mal-information and, accordingly, has different possible remedies. Each of these facets of information disorder is quite distinct where gender is concerned. To understand how they interact with gender, we need to define the relationship between key concepts and terms. Before I explore gendered disinformation through a VAWP lens in the next section, let me briefly discuss how different forms of information disorder interact with gender.

Gender bias is the oil that greases the wheels of misinformation: people are more likely to believe in and spread gendered misinformation if it conforms to existing gender biases. In contrast, gendered disinformation is the willful and malicious propagation and reinforcement of those biases in the first place. For example, a "disinformer" might start a malicious rumor accusing a successful woman of "sleeping her way to the top" (i.e., suggesting that women could not be successful based on professional merit or hard work and/or that women leaders use their sexuality to manipulate). Examples of this exist in every order of the planet: Japanese MP Mari Yasuda has been simultaneously accused of using romantic or sexual affiliation with "powerful men" to progress political career, while being propositioned by the public.[18] This is echoed by the experiences of United States Vice President Kamala Harris, who was accused on a national news program as having "... slep[t] [her] way to the top ...".[19] Individuals might be more likely to believe such disinformation and re-share it online because it aligns with their pre-existing bias about women in politics. This manifests through perceptions that women, by docile "nature", are ill-equipped to lead in political circumstances, and women who do show leadership potential through decisiveness are derided for being less feminine and thus unlikable and untrustworthy.[20] This is compounded across multiracial societies, where women of color are disregarded as "too abrasive" or "too docile" on racial lines, which inherently diminishes qualifications in ways that are not applied to their male political counterparts.[21] In this case, the individual would be sharing misinformation because they believe it to be true.

Gender also operates on mal-information. Women politicians are often especially vulnerable to malicious sharing of personal information. For example, women politicians in multiple countries have been publicly mocked and/or criticized after personal photos are shared of them on vacation (often in informal attire or bathing suits) or together with a past romantic partner. Other women have expressed concern after their personal details (home address, children's school address, etc.) have been shared online for malicious purposes. Women are especially vulnerable to mal-information for, among others, two reasons. One, they are often held to a different standard for moral behavior and/or physical appearance than men, and photos in relaxed settings, with old lovers, or in bathing suits open them to accusations of immoral behavior. Second, they are vulnerable because they (or their children) may be more at risk of physical aggression when this kind of information is publicly shared with malicious intention. These instances of "doxing", defined by Anderson & Wood (2021) as "... a range of acts in which private ... information is published on the internet against a party's wishes",[22] present an opportunity to perpetrators to expose information like an address or phone number to encourage further harassment. In all three cases, digital technology – especially social media – facilitates rapid and widespread diffusion of these information disorders (Figure 7.1).

Looking at gendered disinformation through a VAWP lens

Before we dive into the important but niche area of gendered information disorder, let's take a moment to understand why gendered disinformation and VAWP are so commonly associated. I argue that this occurs for four main reasons:

1. symbiosis exists between harmful stereotypes and bias, disinformation and VAWP,
2. disinformation is a form of psychological harm, which is the most common form of VAWP,
3. the term "symbolic violence" has been used in relation with VAWP when, in fact, it is better suited to explain information disorder and
4. the location of the harm is mistakenly used as the primary classification.

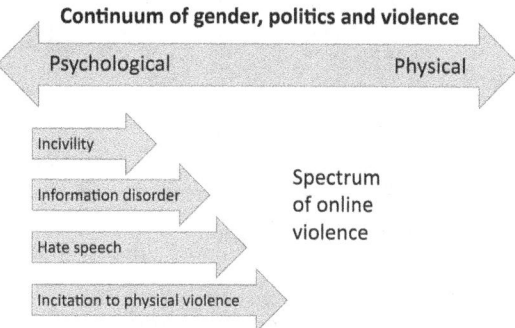

FIGURE 7.1 Locates Gendered Information Disorder within the Broad Universe of VAWP and Recognizes Gendered Disinformation as Just One Form of Information Disorder

Source: Bardall in Zetterberg & Bjarnegård (2022)

First, tactics of political malpractice in the digital sphere fall roughly into two categories. First, "hard" tactics coercively disrupt online or computer-based systems and infrastructure to achieve political advantage or gain (i.e., cyberattacks, election hacking, etc.). Second, "soft" tactics of political malpractice that target values and culture to amplify social division. The latter is achieved through tactics such as false and/or malicious distortions of information and attacks on protected identities. Both types of tactics can be used by bad actors to influence political outcomes, either domestically or abroad. Gender is one of many issues that are commonly leveraged as part of soft tactics in digital political malpractice.

On the other hand, there are multiple drivers of political gender inequality. These range from structural inequalities such as economic disparity and lower social status, to legalized discrimination and discriminatory practices in politics (old boy networks, political money networks, etc.). Harmful stereotypes and biases are major factors contributing to this inequality, worldwide. VAWP is the weaponization of these biases to achieve political outcomes. As depicted in Figure 7.2, information disorder is the inflection point between existing stereotypes, biases, and active harm.

This synergic relationship between political malpractice and political gender inequality is the first reason for the common association between VAWP and disinformation. Understanding this relationship enables us to better situate the discrete challenges involved in each.

Second, there is a still deeper relationship between VAWP and gendered disinformation. Gendered disinforming is, at its roots, a form of gender-based psychological violence. It is used to shame, intimidate, demean, embarrass, harass, ridicule and/or gaslight women *because* they are women – all classic forms of GBV. Research recognizes that psychological violence is by far the most common form of VAWP and that social media offers suitable conditions for psychological violence to flourish.[23]

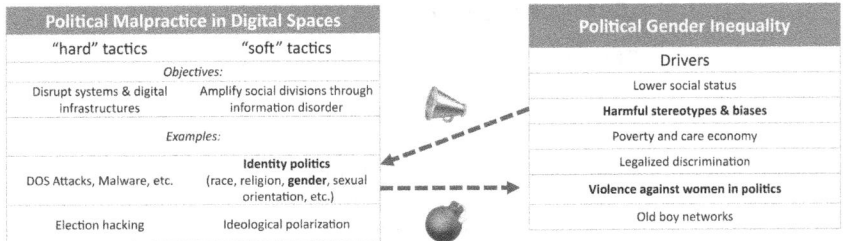

*Existing biases & stereotypes become
forms of VAWP when they are
weaponized through information
disorder*

FIGURE 7.2 Information Disorder Is the Inflection Point between Existing
Stereotypes, Biases, and Active Harm

Globally, non-physical, socio-psychological forms of violence are the most common forms of harm directed at women in politics, both online and offline, in contrast to men who experience (and perpetrate) more physical violence.[24] Non-physical forms of violence are the most widespread and are equally damaging as physical and sexual forms of violence. As discussed below, spreading gendered disinformation against a political woman, which may include defamation, would be classified as a form of socio-psychological VAWP.

Third, sociological forms of violence are also sometimes identified in relation to VAWP, including symbolic or semiotic acts of violence. Comprising acts which "... delegitimize female politicians through gendered tropes denying them competence in the political sphere ...", symbolic violence "... operates at the level of portrayal and representation, seeking to erase or nullify women's presence in political office".[25] I find limited use for "symbolic violence" as a category of VAWP[26]; however, the classic sociological theory of symbolic violence can help explain malicious information operations for two reasons:

- As conceptualized by Bourdieu, symbolic violence relies on the perception of legitimacy and the subsequent consent and complicity of actors concerned.[27] To demonstrate symbolic violence in relation to disinformation: in the case of foreign disinformation, the "dominators" are states like Russia and China, and the "dominated" are the information consumers in the receiving country that consume, endorse, spread and amplify their messages. In such cases, coercion occurs when the dominated are knowing or unknowingly subject to domination because they assume the situation is normal, legal and legitimate.[28] The dominated (wittingly or unwittingly) comply with or perpetuate violence when they accept and amplify gendered disinformation because it conforms to their pre-existing biases, regardless of whether the information is false. This unconscious complicity between dominated and dominator is the defining characteristic of symbolic violence.

- The other connection between symbolic violence and gendered disinformation relates to the nature of the harm that is caused. Common forms of online and offline VAWP are destructive and functional, for example, forcing a woman to resign or drop out of a campaign. They tend to violate social norms and laws and are often interpersonal. However, like Bourdieu's symbolic violence, some disinformation campaigns are better understood as a generative form of violence. Because they build on and amplify existing bias and stereotypes, they serve as "a mechanism to constitute, uphold and organize existing social relations".[29] While these categories are not absolute (i.e., some disinformation campaigns target individual women with the goal of forcing them out of politics), they provide general guidelines for understanding how these different forms of harm operate.

In sum: the term "symbolic violence" has been misapplied to VAWP when, in fact, it is a much more useful concept to understanding gendered disinformation campaigns. The misuse of the term has contributed to confusion around the relationship between VAWP and disinformation.

Finally, the fourth reason that gendered disinformation and VAWP are so commonly associated is because of *where* it most often takes place. More often than not, the backlash against political women that we call VAWP occurs on digital platforms.[30] Online vitriol directed at women in public life, because of their gender, is an intentional form of GBV that aims to control, coerce or silence their political participation. It often builds on broader patterns of gendered exclusion, discrimination and violence. As a form of GBV, the harms of online abuse and violence are gendered in significant ways, such as driving women out of political spaces and contributing to maintaining patriarchal political institutions and policies. As much as the digital space facilitates the fostering of activism against gendered violence, such as with community organizing and discourse in the aftermath of the murder of Sarah Everard,[31] the perceived disconnect from offline impact encourages scaled, publicly-facing sexist harassment through "… anonymity and unaccountability in a disinhibiting environment".[32] Digital platforms lend themselves to this kind of misuse for various reasons:

- they have low-cost barriers to entry;
- users often benefit from impunity due to ease of anonymity, weak legal protections and extra-jurisdictional offenses;
- the speed and breadth at which harm is multiplied, including cross-national bandwagoning and
- toxic "influencers" have outsized impact.[33]

On the other hand, disinformation – and other forms of information disorder – is not technically limited to social media or other online spaces. Yet, given the nature of information sharing today, it is safe to say that the vast majority of disinformation campaigns occur in digital spaces to such an extent as to be effectively synony-mous.[34] The confusion occurs when VAWP and disinformation are associated with each other based on the simple fact that they both often take place online.

Comprised of abusive and threatening messages, memes, videos, GIFs and other web-enabled content as well as hostile digital-based acts like flaming, trolling, doxing, phishing, impersonating and cat-fishing,[35] online VAWP generates vast amounts of wide-ranging online harms across a multitude of platforms and fora, sometimes spilling into offline spaces. Online VAWP is a wildly more complex and multifaceted problem than gendered disinformation. Limiting our interpretation of it to gendered disinformation or information disorder fails to do justice to the many forms of harm that exist. Likewise, trying to classify the broad-ranging harms that political women experience online under the narrow hat of gendered disinforma-tion is inaccurate and ultimately dilutes the concept of disinformation.

Responding to misconceptions about these conceptual cousins

"Disinformation" is a newly-popularized term for an old problem in politics. This final section of this chapter looks at how we can respond to some of these misconceptions, untangle concepts and work with terms that are the most useful for combatting these destructive and harmful attacks on women and on democ-racy in the world.

"Gendered Disinformation" is usually a misnomer

Disinformation amplifies existing bias. It is a downward spiral – harmful gen-der bias and prejudice already exist in societies, so the purveyors of disinfor-mation play on this by spreading false claims about women in politics that are rooted in these biases. People believe the rumors which reinforce their bias, resulting in further exclusion of women from political leadership. For exam-ple, a common stereotype holds that women in politics are not strong enough to lead.[36] A typical disinformation campaign can play on this existing bias, perhaps by manipulating a video to show a woman leader crying or taking real footage out of context. Voters' existing prejudice is thus confirmed and they are even less likely to support a female political leader. Along these lines, according to Barbanchon and Sauvagnat's (2021) study in the French context, in electoral districts where research participants were more likely to agree with statements such as "men are better political leaders than women", it was less likely that women would run for office.[37]

So, if we take this approach to gendered disinformation, simply *holding* a harmful bias or expressing a bias on social media is not the concern. Rather, it is about the act (the lie, the rumor, the manipulated images, etc.) that is used to intentionally and deceptively reinforce an existing harmful gender bias. Bias, hate and misogyny are not disinformation or fake news. They are real. In this sense, the harmful attitudes which perpetuate VAWP are manifested through decisive and malicious opportunities to shift public perception of women. This is done to cater to existing societal norms and public misogyny that lingers in the face of feminist activism and education. In this sense, perpetrators can tap into negative perception and perpetuate disinformation at scale through the digital space. This shifts gendered disinformation from the conceptual to a vehicle for hatred and violence. In this sense, we need to change our thinking to view "disinformation" (or disinforming) as a verb not a noun.

This is a useful way to look at this issue, and it helps separate disinformation from other forms of VAWP. Using this approach excludes many other forms of harassment and abuse (i.e., calling a woman leader a bitch, wishing she gets raped, etc. – these are forms of VAWP but they are not disinformation). They are different, which is important for analysis and response. Looking at gendered disinformation in this way can allow for a legal footing when targeted at an individual woman (see discussion below on defamation). Thus, we might say that "gendered disinformation" is the willful and malicious spreading of false information, based on harmful gendered stereotypes. This is the challenge: "Gendered disinforming" is one means of weaponizing information to perpetrate VAWP.

Just because it's nasty, fake and online doesn't mean it is disinformation

As we have seen, there is a tendency to fold all forms of offensive online behavior against political women under the umbrella of disinformation. Doing so is convenient, but does not ultimately serve the purpose of defining and implementing mitigation and prevention strategies. Some acts of online abuse are easily distinguished from disinformation. This includes acts such as phishing, trolling and doxing, which are malicious but are not clearly related to shifting public perception through information distortion. Others are generally conflated. Too often, we see an expression of gender bias on social media and call it "gendered disinformation". For example, in current writing on gendered disinformation, it is not unusual to see online threats, slurs, insults and degrading humor classified as forms of gendered disinformation.

We need to make sure that the medium does not define the problem here. As infuriating as it may be to some, when an individual sincerely expresses their belief that women in politics are less capable, less qualified, sexually immoral, etc., it is not disinformation per se. It is an expression of personal prejudice or bias. Instead, it is the individual, intentional acts of spreading false information to amplify those biases towards the goal of excluding women from politics that is the actual malicious gendered disinformation. While this may seem like a minor distinction, it has major implications for data collection and for pinpointing "toxic influencers" – those who use their positions of influence on social media to maliciously disinform against women in politics. Once we make this distinction, actors will be more empowered to engage prevention and mitigation actions.

Gendered disinforming in politics may target the public at large, not only the individual or group

VAWP may assume specific relations for the target/victim, as well as for the perpetrator. The target/victim can be a specific individual or group of individuals, with shared values or affiliation beyond gender, including women candidates or MPs, political activists, civil society leaders, electoral officials or journalists. The perpetrators of VAWP are equally identifiable individuals and can include members of opposing political parties or even family members and traditional community leaders.

This relationship often becomes murky online, where anonymous users run rife and bandwagon effects are common.[38] However, unlike other forms of psychological VAWP, gendered disinforming may target the public at large rather than focusing on the targeted woman only. Specifically, although gendered disinforming frequently attacks an individual woman by name, and has many direct negative consequences on the woman under attack, the ultimate objective of the disinforming is to influence the public perception of women in politics. To provide a tangible example: the instigators of the social media campaign to depict former Canadian Environment Minister Catherine McKenna as unintelligent/an airhead could be described as active "disinformers" because they leveraged gender bias and common tropes to undermine a woman leader. Yet, many (perhaps most) of the thousands of toxic messages involved were not directly sent to Min. McKenna at all and were likely never seen by her or her staff – they were shared on personal Twitter feeds of conservative media groups.[39] The "disinformers'" primary audience to consume the disinformation was not Min. McKenna; rather, it was the general public. By hardening existing sexist bias and amplifying harmful stereotypes about women in politics, the disinformers contributed to the deepening of harmful social norms.

Likewise, there may be no direct contact between the perpetrator and the victim. In other cases of online abuse, such as trolling, abusers tag the victim or post directly on their social media spaces. In the case of gendered disinforming, the perpetrator may never seek to directly contact the object of their abuse. The victim may be unaware of the disinformation for an extended time or unaware entirely. This is also an important distinction to make in connecting gendered disinforming to VAWP because it impacts how incidents are identified and measured and the responses that are adapted to resolve them.

Gendered disinforming can be a mobilization tactic of collective violence

It is useful to explore some tactics of gendered disinforming as a vehicle of collective violence. Collective violence is defined by the World Health Organization (WHO) as the instrumental use of violence by people who identify themselves as members of a group against another group or a set of individuals, in order to achieve political, economic or social objectives.[40] The object of collective forms of VAWP is to maintain politics as a patriarchal space. This is to say, by perpetuating falsehoods about women in public spaces, there is less opportunity for these women, and their future counterparts, to access any societal power or bargaining opportunity. To limit the opportunities of women and maintain power structures for men, perpetrators could discredit women based on a shared characteristic that cannot be extended to their male counterparts: attacking their gender. Using disinformation to mobilize others to this end may instrumentalize that information as a part of a collective act of violence. In the case of gendered disinforming, the weaponization of information can be used as a tool to enflame others to action. The actions the others take vary broadly, from trolling to sharing degrading memes to attending protests to ultimately seeking to silence and marginalize a political woman. As a whole, they constitute collective acts of VAWP.

This is a useful frame because there are already multiple legal definitions to capture interpersonal acts of gendered disinformation. These consist of primarily of defamation laws. Defamation refers to harming another person's reputation by making a false written or oral statement about that person to a third party.[41] When a permanent record exists, such as an email, YouTube video, Tweet, Facebook post or blog post, it is called libel. When such acts consist of spoken statements or gestures, they are referred to as slander. Many countries have well-developed defamation laws to address cases such as this. Advancing gendered disinformation as a form of interpersonal violence would be duplicative. Instead, understanding gender as an aggravating factor in cases of defamation of women in politics would be helpful to taking full measure of the harm caused by these acts.

Hostile foreign meddlers are engaged in gender disinforming against women in politics

The final issue is the question of authoritarian strategy. Disinformation can be a tool of state powers (both foreign and domestic) as well as non-state actors and extremist groups. In the intense focus on the content and the methods of this kind of abuse, the role of hostile foreign actors or authoritarian domestic movements is often overlooked. We need to better understand why state actors would engage in VAWP and gender disinforming.

Recognizing that states are male-dominated and inherently misogynist is not enough to explain why state intelligence agencies would launch covert disinformation campaigns. Although the rise in women in politics in past decades is remarkable, it is not enough to interpret gendered disinformation as a coherent backlash. Nor can we presume that foreign states like Russia or China care enough about keeping American (or Ukrainian or Zimbabwean) women out of power as to focus their foreign intelligence resources on this. And yet, we know that foreign influencers have done just that in those three countries and many more.[42]

The key trait of gendered disinformation is that it aligns with broader authoritarian projects.[43] Looking at this solely as a distinct form of VAWP punts the problem. Gendered disinformation is indeed a "hostile act of tactical political subversion".[44] Why? Because undermining gender equality and promoting sexism are effective tools for autocrats – they build social intolerance, fear of change and the sense of insecurity that autocracy thrives on. In harnessing inherent misogyny, which is often further fostered by autocratic activities and encouraged gender norms by the state, there is an opportunity to diminish, disregard and de-platform women who support political opposition.[45] Assaults on women's rights and promoting all forms of intolerance (including homophobia, transphobia, xenophobia, racism and the like) pave the way for further autocratization.

By amplifying biases by spreading explicitly and intentionally false information that will result in the deeper repression and exclusion of women from political leadership, authoritarian actors advance their aims of creating fear, confusion and division in democratic states. They weaken the fabric of democracy as a whole.

Conclusion

The harmful gendered consequences of information disorders are important but very distinct forms of VAWP. In order to use the concepts most effectively, we need to deconstruct the key assumptions about disinformation. First, we must situate gendered information disorder within the framework of VAWP. Gendered information disorders are a form of psychological VAWP occurring mostly online that appear between low-intensity incivility and overt hate speech on the spectrum of online VAWP. Second, we need to stop using the term "disinformation" as a catch-all and instead understand the different causes, manifestations, consequences and remedies of different forms of information disorder. Third, the term "gendered disinformation" is misleading and should be replaced by an active verb, such as "disinforming". Fourth, we must not let the medium define the problem. Much (perhaps most) of the harm that political women experience online is not disinformation. It is other forms of abuse, expressions of personal prejudice or trolling. Next, gendered disinforming can be directed at the general public and used as a mobilizing tool for collective violence, instead of interpersonal. Finally, gendered disinforming can serve as a tactic of authoritarian influence by both domestic and foreign actors and future studies should invest more time in understanding the relation between authoritarianism and gendered disinforming.

Notes

1 Cardy, Meghan. 2018. "'Lock Her Up': Harassment and Violence against Women in Alberta Politics." *Political Science Undergraduate Review* 3 (1): 45–51. https://doi.org/10.29173/psur48; Raney, Tracey, and Cheryl Collier, eds. 2022. *Gender-Based Violence in Canadian Politics in the #MeToo Era 2022*. Toronto, Canada: University of Toronto Press; Tenove, Chris, and Heidi Tworek. 2020. "Online Incivility & Abuse in Canadian Politics." University of British Columbia: Centre for the Study of Democratic Institutions. https://democracy2017.sites.olt.ubc.ca/files/2020/10/Trolled_

Oct-28.pdf; Wagner, Angelia. 2020. "Tolerating the Trolls? Gendered Perceptions of Online Harassment of Politicians in Canada." *Feminist Media Studies*, April, 1–16. https://doi.org/10.1080/14680777.2020.1749691

2 Rheault, Ludovic, Erica Rayment, and Andreea Musulan. 2019. "Politicians in the Line of Fire: Incivility and the Treatment of Women on Social Media." *Research & Politics* 6 (1): 205316801881622. https://doi.org/10.1177/2053168018816228

3 Ibid; Tromble, Rebekah, and Karin Koole. 2020. "She Belongs in the Kitchen, Not in Congress? Political Engagement and Sexism on Twitter." *Journal of Applied Journalism & Media Studies* 9 (2): 191–214. https://doi.org/10.1386/ajms_00022_1; Oates, Sarah, Olya Gurevich, Christopher Walker, and Lucina Di Meco. 2019. "Running While Female: Using AI to Track How Twitter Commentary Disadvantages Women in the 2020 U.S. Primaries." Papers.ssrn.com. Rochester, NY. August 28. https://papers.ssrn.com/sol3/papers.cfm?abstract_id=3444200.

4 Esposito, Eleonora, and Sole Alba Zollo. 2021. "'How Dare You Call Her a Pig, I Know Several Pigs Who Would Be Upset If They Knew' a Multimodal Critical Discursive Approach to Online Misogyny Against UK MPs on YouTube." *Journal of Language Aggression and Conflict* 9 (1): 44–75. https://doi.org/10.1075/jlac.00053. esp.; Ward, Stephen, and Liam McLoughlin. 2020. "Turds, Traitors and Tossers: The Abuse of UK MPs via Twitter." *The Journal of Legislative Studies* 26 (1): 47–73. https://doi.org/10.1080/13572334.2020.1730502.

5 Bigio, Jamille, and Rachel Vogelstein. 2020. "Women under Attack: The Backlash against Female Politicians." *Foreign Affairs* 99: 131. https://heinonline.org/HOL/LandingPage?handle=hein.journals/fora99&div=17&id=&page=.

6 Di Meco, Lucina, and Saskia Brechenmacher. 2020. "Tackling Online Abuse and Disinformation Targeting Women in Politics." Carnegie Endowment for International Peace. November 30. https://carnegieendowment.org/2020/11/30/tackling-online-abuse-and-disinformation-targeting-women-in-politics-pub-83331.

7 Caeiro, Carolina, and Carolina Tchintian. 2021. "Tackling Online Abuse against Women Politicians." Chatham House – International Affairs Think Tank. November 2. https://www.chathamhouse.org/publications/the-world-today/2021-10/tackling-online-abuse-against-women-politicians.

8 Arya, Divya. 2013. "Why Are Indian Women Being Attacked on Social Media?" *BBC News*, May 8, sec. India. https://www.bbc.com/news/world-asia-india-22378366.

9 *BBC News*. 2021. "MP Flick Drummond Quits Twitter over 'Out of Hand' Abuse," April 21, sec. Hampshire & Isle of Wight. https://www.bbc.com/news/uk-england-hampshire-56829181.

10 Ballington, Julie, Gabrielle Bardall, and Gabriella Borovsky. 2017. "Preventing Violence against Women in Elections: A Programming Guide." UN Women. https://www.unwomen.org/en/digital-library/publications/2017/11/preventing-violence-against-women-in-elections.

11 Bardall, Gabrielle. 2022. "Policy Responses to Gender-Based Political Violence Online." In *Gender and Violence against Political Actors*, edited by Elin Bjarnegård and Par Zetterberg. Temple University Press.

12 Southern, Rosalynd, and Emily Harmer. 2019. "Twitter, Incivility and 'Everyday' Gendered Othering: An Analysis of Tweets Sent to UK Members of Parliament." *Social Science Computer Review* 39 (2): 259–275. https://doi.org/10.1177/0894439319865519.

13 Barker, Kim, and Olga Jurasz. 2018. *Online Misogyny as Hate Crime. A Challenge for Legal Regulation?* Andover: Routledge Ltd.

14 American Library Association. 2017. "Hate Speech and Hate Crime", February 28. https://www.ala.org/advocacy/intfreedom/hate

15 Wardle, Claire, and Hossein Derakhshan. 2014. "Information Disorder: Toward an Interdisciplinary Framework for Research and Policy Making." COE. https://edoc.coe.int/en/media/7495-information-disorder-toward-an-interdisciplinary-framework-for-research-and-policy-making.html.

16 Ibid.
17 Ibid.
18 McCurry, Justin. 2021. "'It Is Bullying, Pure and Simple': Being a Woman in Japanese Politics." *The Guardian*. October 27. https://www.theguardian.com/world/2021/oct/27/being-a-woman-in-japanese-politics.
19 Heil, Emily. 2019. "Tomi Lahren Apologizes after Saying Kamala Harris Slept Her Way to the Top." *Washington Post*, August 1. https://www.washingtonpost.com/arts-entertainment/2019/08/01/tomi-lahren-apologizes-after-saying-kamala-harris-slept-her-way-top/.
20 Kantar. 2020. "The Reykjavik Index for Leadership 2020/2021." Www.kantar.com. https://www.kantar.com/campaigns/reykjavik-index.
21 Ro, Christine. 2021. "Why Do We still Distrust Women Leaders?" *BBC Worklife* (UK), January 19. https://www.bbc.com/worklife/article/20210108-why-do-we-still-distrust-women-leaders.
22 Anderson, Briony, and Mark A. Wood. 2021. "Doxxing: A Scoping Review and Typology." In *The Emerald International Handbook of Technology-Facilitated Violence and Abuse,* edited by Jane Bailey, Asher Flynn and Nicola Henry. West Yorkshire: Emerald Publishing Limited, pp. 205–225. https://www.emerald.com/insight/content/doi/10.1108/978-1-83982-848-520211015/full/html.
23 Bardall, Gabrielle. 2011. "Breaking the Mold: Understanding Gender and Electoral Violence." International Foundation for Electoral Systems. https://www.ifes.org/sites/default/files/gender_and_electoral_violence_2011.pdf; Bardall, Gabrielle. 2013. "Gender-Specific Election Violence: The Role of Information and Communication Technologies. *Stability: International Journal of Security and Development* 2 (3), p.Art. 60. http://doi.org/10.5334/sta.cs; Krook, Mona Lena, and Juliana Restrepo Sanín. 2019. "The Cost of Doing Politics? Analyzing Violence and Harassment against Female Politicians." *Perspectives on Politics* 18 (3): 1–16. https://doi.org/10.1017/s1537592719001397.
24 Bardall 2011.
25 Krook, Mona Lena, and Juliana Restrepo Sanin. 2016. "Violence against Women in Politics: A Defense of the Concept." *Politica Y Gobierno* 23 (2): 459–490.
26 See Bardall, Gabrielle. 2019. "Symbolic Violence as a Form of Violence against Women in Politics: A Critical Examination." *Revista Mexicana de Ciencias Políticas Y Sociales* 65 (238). https://doi.org/10.22201/fcpys.2448492xe.2020.238.68152.
27 Bourdieu, Pierre. 1979. "Symbolic Power." *Critique of Anthropology* 4 (13–14): 77–85. https://doi.org/10.1177/0308275X7900401307. See also, Bardall 2019.
28 Bourdieu, Pierre. 2001. "Television." *European Review* 9 (3): 245–256. https://doi.org/10.1017/S1062798701000230. See also Bardall 2019.
29 Ibid; Colaguori, Claudio. 2010. "Symbolic Violence and the Violation of Human Rights: Continuing the Sociological Critique of Domination." *International Journal of Criminology and Sociological Theory* 3 (2): 388–400, p. 392. See also: Bardall 2019.
30 Bardall 2013.
31 Humphreys, Rachel. 2021. "Why Are Women So Angry after the Killing of Sarah Everard?" *The Guardian*, March 16, sec. News. https://www.theguardian.com/news/audio/2021/mar/16/how-sarah-everards-killing-has-reignited-the-debate-around-womens-safety.
32 Lewis, Ruth, Michael Rowe, and Clare Wiper. 2016. "Online Abuse of Feminists as an Emerging Form of Violence against Women and Girls." *British Journal of Criminology* 57 (6): azw073. https://doi.org/10.1093/bjc/azw073.
33 Bardall 2013.
34 For example, disinformation toolkits from government sources such as the European Commission, the UK's Government Communication Service, ASEAN and USAID as well as major NGOs such as Interaction and the RAND Corporation refer exclusively to online information sharing, signaling that disinformation is an overwhelmingly online phenomenon.

35 The vocabulary of online abuse is sprawling and colorful. Helpful glossaries are offered by the Women's Media Center, https://womensmediacenter.com/speech-project/online-abuse-101

36 Kantar, *The Reykjavik Index.*

37 Le Barbanchon, Thomas, and Julien Sauvagnat. 2021. "Electoral Competition, Voter Bias, and Women in Politics." *Journal of the European Economic Association,* July. https://doi.org/10.1093/jeea/jvab028.

38 Nadim, Marjan, and Audun Fladmoe. 2019. "Silencing Women? Gender and Online Harassment." *Social Science Computer Review,* July, 089443931986551. https://doi.org/10.1177/0894439319865518.

39 Tayler Young. 2021. Thesis. University of Alberta. Forthcoming.

40 World Health Organization. 2002. "World Report on Violence and Health: Summary." Geneva: WHO. https://www.who.int/violence_injury_prevention/violence/world_report/en/summary_en.pdf.

41 Canadian Journalists for Free Expression. 2015. "Defamation, Libel and Slander: What Are My Rights to Free Expression?" CJFE | Canadian Journalists for Free Expression. https://www.cjfe.org/defamation_libel_and_slander_what_are_my_rights_to_free_expression.

42 Bardall, Gabrielle. 2019a. "Autocrats Use Feminism to Undermine Democracy." Policy Options. October. https://policyoptions.irpp.org/magazines/october-2019/autocrats-use-feminism-to-undermine-democracy/; Chenoweth, Erica, and Zoe Marks. 2022. "Revenge of the Patriarchs." Www.foreignaffairs.com. February 13. https://www.foreignaffairs.com/articles/china/2022-02-08/women-rights-revenge-patriarchs?utm_medium=newsletters&utm_source=twofa&utm_campaign=Revenge+of+the+Patriarchs&utm_content=20220211&utm_term=FA+This+Week+-+112017&fbclid=IwAR17cDUiWztgptM4kODK7FrVA2jSd12SmCZT91ESaTjmTmez4kO-YZo5lw0; Donno, Daniela, and Anne-Kathrin Kreft. 2018. "Sometimes Autocrats Strengthen Their Power by Expanding Women's Rights. Here's How That Works." *Washington Post,* November 23. https://www.washingtonpost.com/news/monkey-cage/wp/2018/11/23/sometimes-autocrats-strengthen-their-power-by-expanding-womens-rights-heres-how-that-works/; Thornton, Laura. 2021. "How Authoritarians Use Gender as a Weapon." *Washington Post,* June 7. https://www.washingtonpost.com/opinions/2021/06/07/how-authoritarians-use-gender-weapon/.

43 Jankowicz, Nina, Jillian Hunchak, and Alexandra Pavliuc. 2021. "Malign Creativity: How Gender, Sex, and Lies Are Weaponized against Women Online | Wilson Center." Www.wilsoncenter.org. Wilson Center, January 25. https://www.wilsoncenter.org/publication/malign-creativity-how-gender-sex-and-lies-are-weaponized-against-women-online.

44 https://www.dictionary.com/browse/disinformation

45 Jankowicz et al. 2021.

8

SECURITY, MISOGYNY, AND DISINFORMATION UNDERMINING WOMEN'S LEADERSHIP

Kristina Wilfore

Introduction

"As only the 18th woman to receive this prize, I need to tell you how gendered disinformation is a new threat and is taking a significant toll on the mental health and physical safety of women, girls, trans, and LGBTQ+ people all around the world,"[1] said journalist Maria Ressa in her 2021 Nobel Peace Prize acceptance speech.

Despite a great deal of attention paid to disinformation and its impact on democracy, relatively little consideration has been given to the way in which misogyny intersects with misinformation and extremism online, creating insecure environments ripe for the consolidation of power that leaves women by the wayside. This chapter argues that enhancing a political discourse to acknowledge the role of technology in enabling gendered disinformation attacks on women in politics is not only a priority for securing and protecting women's rights, but also a key foreign policy and national security imperative for democratic-minded countries.

A growing body of evidence shows that authoritarian-minded leaders, whether they are autocrats or malign actors within democratic or semi-democratic systems, are increasingly deploying gendered disinformation campaigns to attack and silence their critics, thereby weakening democracies and their public discourse.[2] Vladimir Putin in Russia, Rodrigo Duterte in the Philippines, Viktor Orban in Hungary, and Recep Tayyip Erdogan in Turkey are among the political leaders who have used gendered disinformation to attack women in politics, aggressively challenge feminism, and undermine liberal values, such as support for civil rights and human rights, pluralism, secularism, inclusion, and freedom of expression. In recent elections in the West, bad actors, as well as locally grown far right movements, have amplified gendered disinformation as a tool in influence operations.

DOI: 10.4324/9781003261605-11

Only by placing gendered disinformation and the fight against authoritarianism at the center of policy agendas to reduce online harms can reform-minded governments score a victory for women's rights, but also advance key national security and foreign policy objectives to prevent further democratic backsliding. Examining the problem of disinformation through a gender lens enhances the societal understanding of how sexist narratives are used to intimidate women in order to eliminate critics and consolidate power, therefore blocking democratic processes.

This deluge of gender-based attacks is not happening by chance but by design, and it will continue to unfold especially in moments of political transition, either in hotly contested elections or when political power is consolidated. Sexism on the campaign trail is, sadly, nothing new. What is new is that digital technology has made it much easier for gendered disinformation campaigns to be organized and amplified, and cheaply financed, resulting in mainstreaming attacks. The way digital media platforms are designed is increasing sexism in politics, as image-based, fact-void content often becomes sticky and viral, with storylines picked up by mainstream traditional news outlets.

The necessity to act urgently could not be more evident. According to Freedom House, 2021 marked the 15th consecutive year of decline in global freedom with the countries experiencing deterioration outnumbering those that improved by the largest margin recorded since the negative trend began in 2006. Democracy defenders have sustained heavy losses in their recent struggle against authoritarians, "shifting the international balance in favor of tyranny."[3] Yet, within security and technology, little attention is paid to the systematic attacks on women leaders as a tactic to undermine democracy and consolidate power when examining antidemocratic global trends and interventions. Therefore, although these are not the only factors at play, the role of the information environment which enables authoritarian tactics, and the digital ecosystem in which bad actors take attacks to scale, must be contended with seriously in this context.

This chapter begins by defining the problem of gendered disinformation and describing the typology of narrative tactics that intersect with gender bias. Next, it expands upon the role that digital platforms play in gendered disinformation campaigns. The chapter ends with an outline of possible countermeasures, such as establishing global frameworks against gendered disinformation and new social media standards, and building digital resilience.

Defining the problem: Gendered disinformation and women leaders

Lucina Di Meco (2020) defines gendered disinformation as the spread of deceptive or inaccurate information and images used against women political leaders, journalists, and female public figures, following storylines that draw on misogyny and stereotypical gender roles.[4] Building on sexist narratives that are characterized by malign intent and coordination, gendered disinformation both distorts public understanding of female politicians' track records and discourages women from seeking political careers.[5] In another notable working definition, Demos (2020) describes gendered disinformation as "information activities (creating, sharing, disseminating content) which attacks or undermines people on the basis of their gender; weaponises gendered narratives to promote political, social or economic objectives."[6]

The definitions are evolving as evidence surfaces about the role of technology in extenuating hate and bias, and the ease in which those tools can be weaponized for malign intent. The concept of gender-based violence online (GBVO) often gets transposed with gendered disinformation, when in fact they mean different things. All of these phenomena fall into the larger frame of gender-based violence (GBV). The United Nations defines violence against women as "any act of gender-based violence that results in, or is likely to result in, physical, sexual, or mental harm or suffering to women, including threats of such acts, coercion or arbitrary deprivation of liberty, whether occurring in public or in private life."[7] Gendered disinformation involves the intentional spread of false information *about* persons or groups based on their gender identity, and often refers to the overall ecosystem in which this information is organized.[8] GBVO and harassment involve *targeting* and abusing individuals based on their gender identity, but may not involve the main features of disinformation (content that is intentionally false *and* designed to cause harm). Social media posts calling women politicians "whores" or "fat pigs," for example, may or may not be defined by digital platforms as GBVO or hate speech, depending on the platforms in which this information is shared and their adopted definitions. Yet, these attacks fall within the broader playbook of gendered disinformation. Such attacks are not provable as false per se, but are intended to do harm, and are a key component of the ecosystem in which extremism is fostered, on and offline, by mis- and disinformation and its inherent relationship to both technology and digital platform design.[9]

Gendered disinformation builds on the combinations of three vectors of threats women leaders face on social media: (1) sexism, misogyny, and the manosphere (a "disparate, conflicting and overlapping men's groups… shar[ing] a hatred of women and strong antifeminism")[10] (2) online violence (threats, abuse, hate speech, harassment, smear campaigns), and (3) disinformation (malign actors, artificial activity, coordinated influence operations, fake news).[11]

Many gender advocates working in this space approach these issues through the Violence Against Women in Elections (VAWIE) framework, describing online attacks as the result of misogyny and patriarchal social norms, and focus on supporting women to find individual protection from online harms. This understanding has, however, important limitations, as it fails to address how and why these attacks on women politicians have become so dangerous and pervasive – such as through algorithmic preferences that privilege fake and outrageous content as it enhances profit.

Typology of narrative tactics and the intersection of gender bias

In order to both understand gendered disinformation and work toward meaning-ful solutions to weaken its impact, one must first appreciate how pervasive sexism in politics is, and how it is manifested during campaigns, on and offline, and throughout the period in which women hold public office. Gender-based attacks online are a highly concerning development for society at large, and worthy of more scrutiny and response, as attacks meant to silence or undermine women in politics are attacks on democracy and equal representation. Often, the goal of these attacks is to spread false information, sow hatred and agitation toward women, and signal to society that women do not belong in powerful positions of leadership.[12]

Bias toward women leaders is exploited through gendered disinformation campaigns.[13] The expression of bias in modern political campaigns has been researched over many election cycles by the Barbara Lee Family Foundation. Their findings demonstrate that women candidates need to meet a higher stand-ard for qualifications compared to male candidates, be more likeable to be seen as qualified, and are judged more harshly on their appearance, as well as related to their family responsibilities.

- **Higher standards for proving qualifications.** Disinformation plays into societal bias against women, who throughout history have had to prove their qualifications to serve, while the qualifications of men are always assumed by a public that is more forgiving for any wrongdoing.[14] The reality is that candidates are imperfect human beings and questions about qualifications are a natural part of campaigning. Yet, for women candidates, mistakes become outsized sources of attack about their fit for public office.

  The 2021 German federal election for chancellor illustrates this point very clearly. Since the beginning of Green Party candidate Anna Baerbock's candidacy, she faced a disproportionate array of vicious online attacks from both foreign and domestic actors, often steeped in sexism. According to a study by the Institute for Strategic Dialogue, 18 of the most shared Facebook posts about Baerbock contained false information or allusions to conspir-acy theories, compared to only three of the posts about Olaf Scholz and none of the posts concerning Armin Laschet. This trend was confirmed on Telegram, where 43 of the most popular posts about Baerbock contained misinformation or conspiratory narratives, compared to only 26 posts about Laschet and 17 of the posts about Scholz.[15]

  Not only was Baerbock attacked more often than her male competitors: the nature of the attacks against her was different too, as it was more personal (often referring to her "incompetence" as opposed to her policy proposals or ideas), hyperbolic (on Facebook, Baerbock was described 15 times more often than Scholz and 7.5 times more often than Laschet as a "danger to Germany"), and most often sexist, containing clear references to her gender.

- **Appearance-based attacks.** No matter the level of office, women in politics are subjected to more undue criticism of their appearance, voice, and clothing in comparison to men, which has a direct impact on perceptions of their experience, with the subtext that women are not as fit for office.[16] According to research conducted by GQR, a public opinion and data analytics firm, a tactic among misogynistic social media influencers in the United States driving attacks is to feed Instagram accounts with harmful images that mock women leaders for their appearance, or edit images of male celebrities and politicians to make them look like women which are picked up and utilized by far right influencers.[7] Such tactics do not violate digital platform policies but are a pervasive part of the gendered disinformation formula used by bad actors.

  In Canada, for example, women politicians and journalists are very concerned that online harms are having a chilling effect on women's political and civic engagement. Former Canadian environmental minister Catherine McKenna was the target of a massive online campaign of gendered abuse which referred to her as "climate Barbie." She also faced sexualized insults and threats against her family, to the point that Ms. McKenna was assigned extra protection and a security detail – measures that are quite exceptional in the Canadian political context. She recently resigned from her post.[17]

- **Higher standards of likability.** On top of demonstrating competence, women face a higher threshold to be liked by voters. According to public opinion pollster Celinda Lake, voters will support a male candidate they do not like but believe is qualified, whereas they will not support women who are considered unlikable.[18] Women pollsters and public political researchers contend that likability of candidates and elected leaders, is a highly subjective political concept where issues such as a woman's tone of voice, appearance, and demeanor are factored heavily in this assessment.[19] These personal characteristics become trigger points of gendered disinformation with a voluminous amount of social media content focused on such attributes in order to attack and demean women. Even women judge women candidates more heavily on all these factors.[20] Further compounding this dynamic, evidence shows that when women candidates break through the likability barrier, voters show hesitancy to support a woman because they think other voters aren't ready for a female candidate or won't like them as much compared to men.[21]

- **Threats that are more violent, racists, and sexist.** A growing body of evidence shows that once elected, women and people of color continue to bear the brunt of harassment and threats at all levels of government,[22] which is compounded further for Black women who experience both systemic racism and sexism at the same time, known as *misogynoir*.[23] "While criticism of their male colleagues tends to focus on name-calling, women in public-facing roles are more likely to face physical threats," according to Mary Anne Franks, a professor at the University of Miami's School of Law, who researches the intersection of civil rights and technology.[24] Well documented in "The Abuse and Misogynoir Playbook," researchers show how technology and algorithms work to silence, shame, and erase Black women and their contributions to society.[25] This is particularly relevant for women candidates and elected leaders required to engage with wide swaths of society on and offline.

One recent study involving 13 female politicians across three countries; Russia, Iran, and China; and six social media platforms found that nine of them had been targets of gendered disinformation narratives.[26] Further, a study from Demos and the European Union (EU) Disinfo Lab unveiled a vast network of actors, at work to delegitimize women politically, by weaponizing sexist narratives in Europe.[27] A majority of the 88 female politicians and experts interviewed for the 2019 report #ShePersisted Women, Politics & Power in the New Media World reported extreme concern about the pervasiveness of gender-based abuse and disinformation in digital spaces.[28] They described it as a barrier for women who want to engage in politics and a serious disincentive for young women to consider political careers.

In Ukraine and Brazil, women politicians' morality and dignity have been tainted by fake stories and disinformation campaigns. In 2018, following the murder of Marielle Franco, a Brazilian city councilmember and human rights activist, a network of bloggers, prominent businessmen, and lawmakers close to Brazilian President Jair Bolsonaro carried out a gendered disinformation and smear campaign against Franco, claiming she had led an "immoral" life.[29] Franco had been deeply critical of Bolsonaro's language and policies.

As Franco's case demonstrates, these tactics of gendered disinformation have been embraced at the highest levels of politics and are increasingly becoming a centerpiece of authoritarianism globally.[30] State-aligned actors in the Philippines and Poland have spread false, humiliating, and damaging narratives against female politicians aimed at shielding the ruling power, according to a recent study from Demos,[31] and Indian Prime Minister Narendra Modi has followed Twitter accounts responsible for rape and death threats against female politicians in his own government. Modi's party has been accused of running a "troll army," targeting political opponents, especially prominent female figures, with online harassment, abuse, and disinformation campaigns.[32]

- **Double bind around family responsibilities.** Women in the public eye are also scrutinized and judged more harshly when it comes to family responsibilities. Voters recognize a double standard exists for moms, but research shows they actively and consciously participate in it.[33] This becomes particularly challenging for women running for office, who increasingly utilize social media to present themselves as relatable to the public as moms, yet face exploitation by opponents online who utilize sexist tropes to question whether women can truly balance the competing priorities of their families and their constituents, particularly when their children are young.[34]

Participants taking part in 2020 series of digital roundtables in Canada reported witnessing and experiencing a growing amount of gendered online abuse and disinformation, aimed at threatening and dehumanizing them, and undermining their credibility, while sometimes targeting their families.[35] Tamara Taggart, a former journalist and federal parliament candidate, said that she was the target of relentless "gaslighting" and politically motivated trolling, often led by her political opponents, which included attacks calling her a "retard breeder" on social media – a hurtful and inhuman reference to her son's disability.

- **Sexualized attacks meant to humiliate and demean.** Sexualized attacks[36] are a constant backdrop to disinformation aimed at women, with memes and graphics accusing women of "sleeping their way to the top" or being sexually promiscuous. Image-based attacks typically play on both bias in society and are sexual in nature, relying on sexist tropes that women are weak and incompetent. Additionally, accusations that position women as too libidinous or too emotional to hold office or participate in democratic politics are a nonstop, underlying the tone of gendered disinformation.

In Western Europe, far right political actors have used gendered disinformation narratives against progressive female legislators, deeming them unfit to hold political power. A global survey of women parliamentarians from the Inter-Parliamentary Union found that 42% of the respondents had seen extremely humiliating or sexually charged images of them spread through social media, including photomontages showing them nude.[37]

Former President Kolinda Grabar-Kitarovic of Croatia became subject to a gendered disinformation campaign that attempted to undermine her credibility by painting her as a hyper-sexualized figure. In this instance, tabloids ran pictures of another woman in a bikini and falsely claimed it was her. The photo's subject was later identified as Coco Austin, the partner of American rapper Ice-T – but, at that point, the damage to Ms. Grabar-Kitarovic's reputation was smeared by the online chatter associated with the false imagery.[38]

Each of these preexisting biases are the threat surface on which gendered disinformation campaigns are organized and amplified, with the goal of mainstreaming attacks. Through cross-channel repetition, coordinated sharing, and means of simulating artificial topic momentum, attacks to undermine women can easily be taken to scale. In fact, a study undertaken by the Institute for Strategic Dialogue to analyze the scale and nature of online abuse targeting 2020 US Congressional candidates showed that women were far more likely than men to be attacked on Twitter, with abusive messages making up more than 15% of the comments directed at every female candidate analyzed, compared with around 5–10% for male candidates. Further, women of ethnic minority backgrounds were particularly likely to face online abuse.[39]

In sum, gendered disinformation campaigns build on, and are rooted in, deeply set misogynistic frameworks and gender biases that portray masculine characteristics as those fit for leadership while painting women leaders as inherently *untrustworthy* (insinuating a woman is dishonest or can't be trusted is a tried and true attack), *unqualified* (one of the biggest barriers women face when seeking office), *unintelligent* (tropes about women as dimwitted and unfit for the job are a prominent feature of gendered disinformation, made worse by objectifying, sexualized content), and *unlikable* (which for women can be the death knell of their campaign).[6]

Implications of gendered disinformation

Beyond undermining specific candidates or women leaders, state-aligned gendered disinformation campaigns are used as a deliberate tactic to smother opposition voices, erode democratic processes, and silence demands for government accountability.[40] Research has shown that women's political participation often represents a challenge to entrenched illiberal and autocratic political elites, disrupting what are often male-dominated political networks that allow corruption and abuse of power to flourish.[41] That is why gendered disinformation has been used by some governments to silence demands for a change and undermine calls for better governance. Particularly in countries where women are among the most outspoken critics of so-called "machismo populism," gendered disinformation perpetuates the notion of politics as an inherently corrupt, cynical, and violent field, unfit for those who are not willing to play dirty.

Such campaigns can thwart progress on gender equality policy agendas by increasing societal intolerance to GBV protections and inflame culture wars around the LGBTQ+ community and ethnic minorities. There is evidence that women politicians might refrain from espousing feminist views because of the strong possibility of an online backlash from gender trolls, while other women who want to pursue feminist policy aims might opt out of electoral politics to avoid the online harassment that women politicians experience.[42]

Pushing women out of the political arena is often only the first step of a broader, dangerous strategy to erode democracy and human rights. According to the United Nations (UN) Human Rights Council, the erosion of women's human rights "is a litmus test for the human rights standards of the whole of society," and this tech-enabled backlash against women's rights has broader ramifications for global peace and security.[43]

The role that digital platforms play in gendered disinformation campaigns

Online hate, extremism, and targeting of women leaders is both possible with the tools provided by digital platforms, and profitable. Until the business model shifts, gendered disinformation campaigns will continue largely unabated. While reform across the EU, the United Kingdom, Canada, the United States, and Australia continue to be seeded, governments need to strengthen their leverage with the platforms to address gendered disinformation, and demand transparency and the fulfillment of existing terms of service to help reduce harm.

Critical to shifting regulatory approaches is recognizing that social media platforms are not neutral disseminators of information, but curate content to serve their commercial interests, and therefore they must assume greater responsibility for the harms they amplify and spread. The future of democratic discourse, and the ability for women candidates to compete fairly, hinges on accepting this fundamental reality.

Algorithmic preference and extremism

The most hateful content directed at women is amplified by algorithms which reward[44] extreme and dangerous points of view with greater reach and visibility, creating a fertile breeding ground for bias to grow. While sexist attitudes are integral to understanding violent extremism and political violence, social norms per se do not explain how attacks against women in politics have been able to proliferate at such a rapid rate online. Cynically coordinated campaigns by illiberal actors are able to take advantage of algorithmic designs and business models that incentivize fake and outrageous content.

Conventional wisdom points to social media as merely a reflection of societal sexism and misogyny, yet this does not take into consideration the way social media acts as a behavioral modification system that encourages groups of individuals to behave in ways that they would not normally and uses algorithmic principles to manipulate people. Facebook/Meta whistleblower Frances Haugen, for example, recently described to UK and US lawmakers how the company studies its products' failures and then buries results that don't promote the company's interests, including on the health and safety of kids, particularly teenage girls. The evidence surfaced shows that Facebook/Meta actively endangers women and girls on its platforms and consistently contributes to the spread of disinformation that has put democratic processes worldwide at risk.[45]

Platforms use deep learning algorithms designed to prioritize disseminating content with greater engagement potential – regardless of whether it is truthful, or irrespective of harm or social impact.[46] The primary way social media companies like Facebook and Twitter make money is through selling advertising, with profits earned from advertisers based on "monthly active users" (MAU), which is a term that refers to the number of unique customers who interacted with a product or service of a company within a month. Essentially, the MAU is a key performance indicator that is used to gauge the overall performance of a company that measures online user engagement.[47] Engagement is currency for digital platforms. Online harms against women are easily amplified by the social platforms, whose business model benefits from gendered disinformation when it increases engagement and salacious content.[48] The algorithms of these platforms actively play into cyber violence against women politicians and journalists at the expense of social cohesion and inclusivity.

Lack of enforcement of standards

Although companies adopt general terms of services and codes of conduct that ban hate speech, harassment, and the promotion of violence,[49] their implementation has so far been very unsatisfying, due to inadequate and unclear content moderation systems[50] that rely on "notice and take down models," as reports of abuse are handled reactively on a case-by-case basis. The ongoing move to more automated content moderation – explained by many social media companies as a way to address hate speech – is also unlikely to significantly improve things, as these tools have been proven to be often biased and ineffective.

Digital platforms across the board do not consistently remove posts that violate terms of service or punish repeat offenders. For example, Facebook/Meta has policies against threats of violence, hate speech, violent and graphic content, nudity and sexual activity, cruel and insensitive content, manipulated media and deepfakes,[51] fake accounts, and coordinated inauthentic behavior, yet digital forensic analyses, and even unsophisticated searches, demonstrate that gendered disinformation, which violates such policies, is thriving on their platforms.[52]

Countermeasures

Establishing global frameworks against gendered disinformation, developing new social media standards, and building digital resilience are all ingredients of strengthening countermeasures against gendered disinformation. The landscape for targeted countermeasures is extremely fluid. The ideas in this section reflect the available knowledge at the time of writing this chapter.

Establishing new social media standards

In order to address gendered disinformation and online abuse against women in politics, it is essential to understand the incentive structure that allows for this type of content to thrive, and devise regulatory mechanisms for social media platforms that establish better standards for consumers.

Even while recognizing private companies' purview to determine their own business model, governments must create regulatory frameworks that set the stage for better social media standards. Efforts to encourage technology companies to change their products and practices to reduce harm are an enormous undertaking against a powerful, largely unregulated industry. It is essential that those reform efforts move forward with a more diverse set of countries, and are informed by the experience of women and the growing knowledge of how gendered disinformation manifests itself.

As countries are forming regulations to reduce online extremism and hate speech, the reflexive approach of digital platforms, and some critics of this approach (who often take funding from the platforms),[53] is to point to the precedent this may set for authoritarian and illiberal contexts with legal frameworks that do not protect the freedom of expression, as they might be used as tools to undermine political opposition and freedom of speech. The assumption that authoritarian regimes are not already benefiting from existing irresponsible practices of digital platforms is false.[54] Such critics should be equally worried about the freedom of speech of the women who are attacked and silenced online. Young women report being discouraged from seeking leadership roles[55] as a result of this hyped-up environment online, others report self-censoring or totally disengaging from social media, resulting in a chilling effect of freedom of expression for women,[56] including journalists, and particularly minority women. For women who can't afford to disengage because communicating online is part of their job – like women in politics and journalists – the psychological toll on them and their families is incommensurable.

Digital researchers, disinformation experts, and the women's international NGO community in particular need to be savvier in their advocacy around digital regulatory reform to ensure that their recommendations are not gender-blind or less than relevant to a meaningful engagement in mitigating gendered disinformation. When Facebook/Meta, Instagram, Twitter, TikTok, and Google announced at the 2020 UN Generation Equality Forum their commitment[57] to improving the reporting process for women under attack, a core of civil society groups deeply engaged in mitigating the impact of disinformation on women candidates in recent elections, moved swiftly to point out[58] that such insufficient commitments distract from the very epidemic of GBVO that the UN forum was attempting to highlight, and that false solutions are simply victim-blaming by telling women to "cover up" online to prevent their own harassment.

Establishing global frameworks against gendered disinformation

The playbook of coordinated harassment campaigns targeting female identifying politicians and journalists has been unfolding in many countries where women, in particular women of color, are disproportionately targeted by abuse on both Twitter and Facebook/Meta. The deluge of such gender-based attacks is not happening by chance, but by design, and will continue to unfold election after election.

Given the uncertain trajectory of any one piece of legislation meant to curtain online harms, it is important to recognize that there are many reform principles that can address online gender harms, and should be part of the political dialogue around tech policy reform going forward.

Policy principles can be incubated through Transatlantic cooperation. Meaningful juridical interventions in one country (or the European Parliament) to curtail digital distortion can act as a signpost for other policymakers. European regulators have a unique leadership role to play in the three-dimensional chess game of digital policy reform, just as they did with data privacy and protection through the EU's laws on data protection and online privacy codified in its Generalized Data Protection Regulation (GDPR). This will require that regulators sophisticate their analysis to move beyond what is illegal or legal content, and instead channel reform conversation to demand greater transparency and real accountability from digital media companies for the harms that are being caused by the malign use of their products.

Transparency and data-sharing requirements

Transparency and data-sharing requirements of digital platforms are important standards to establish in any meaningful reform effort. Routinely tracking and monitoring harmful content on digital platform sites would raise public awareness about the risks posed by engagement – and with public knowledge comes incentive for platforms to better moderate harmful content and behavior.

A lack of enforcement of regulatory measures from digital platforms means that malign actors can easily take advantage of weak self-regulation structures. Disinformation analysts point to inconsistent policy enforcement or a lack of sanctions for rule-breakers as one of the main problems contributing to an unsafe environment for women, and an uneven playing field for women in politics. Opaque platform data preventing independent oversight or research is also a problem, as well as a reliance on external researchers to flag violations reactively, and, overall, weak supervision and lack of transparency in decision-making make the technical challenges of curtailing disinformation even more complex.

Increasing transparency could enable researchers to better understand metrics of gendered and sexualized harassment and disinformation, including the number of cases that have been reported, who were their perpetrators, and how they were addressed. Such information is essential to understand and evaluate the breadth and magnitude of the investments that social media companies have been making to address this issue (including the number of people working on content moderation, their cultural competency, their position and power within the companies, and their influence on larger decisions, for example, the design of income-generating algorithmic preferences), as well as identify areas for improvement.

Duty of care

Infringing on the freedom of expression is one of the main perceived obstacles to digital policy reform and regulation. As many have noted, free speech is not the same as free reach.[59] While users may have the right to say something inflammatory, false, and offensive online, that should not imply that companies can amplify certain narratives disproportionately and artificially through algorithmic preferences that aim to attack and undermine women candidates and journalists.

The United Kingdom's Online Harms approach demands from large social media companies a statutory "duty of care" requiring them to take action against content or activity that "cause[s] significant physical or psychological harm to individuals."[60] Similarly, the Digital Services Act in the EU envisages the creation of due diligence obligations for platforms' content moderation activities, as well as an obligation for social media companies to conduct risk assessments at least once a year on the systemic risks stemming from the functioning and use of their services, not only as referred to the dissemination of illegal content, but also to the intentional manipulation of their services, through the creation of fake accounts or bots, for example, as well as any negative impact their services might have on the exercise of fundamental rights – including the prohibition of discrimination.

Attention to authoritarian actors

Collectively, prodemocracy leaders must build an architecture of actors and institutions tasked with analyzing how autocratic governments, and authoritarian-like actors in stable democracies, are using social media to undermine women in politics and democracy itself. This architecture can enable the creation of targeted short- and long-term countermeasures to interrupt gendered disinformation.

Resources should be devoted to understanding the misogynistic roots of disinformation, perpetrated by domestic actors but often dovetailing with foreign influence operations or extremist movements. Alarmingly, policymakers, media, and digital investigators are not connecting the dots between extremism and the online targeting of women leaders – despite the growth of violence-prone incel movements and explicitly misogynistic communities such as the Proud Boys. According to the Anti-Defamation League, online misogyny is embraced by white supremacists and other right-wing extremists partly as an organizing tool, to present themselves as defenders of "conservative" values in order to make their views more acceptable to a wider audience.[61]

With a deeper understanding of how this system works, and how online harms against women leaders are intrinsically tied in with the design of social media platforms, we can create greater urgency for demanding new regulatory frameworks to reduce harms while affirming existing security and gender equality commitments embedded in various international treaties, such as the Convention on the Elimination of All Forms of Discrimination against Women (CEDAW), the Beijing Declaration and Platform for Action, and the UN Security Council resolution 1325 on women, peace and security.

Gendered disinformation task forces

Formation of country-specific Gendered Disinformation Task Forces to consider platform accountability and to study harassment, extremism, and violence against women is a reasonable starting point. Such a Task Force could work with academia, think tanks, women's rights organizations, and technologists to map disinformation campaigns and their connection to online abuse of women in politics, pressing for better access to platform data to conduct further analysis. Such a Task Force can ensure that women's perspectives and lived experiences as candidates and elected officials across party lines are represented, especially as new regulatory approaches are devised. With the creation of Task Forces, transatlantic, as well as across the Global North and South, it will be more feasible to share findings and best practices for interventions through a global dialogue that examines the issue comprehensively, and with an eye toward mitigating harm.

Building digital resilience

Women leaders at the forefront of democratization efforts are the targets of disinformation and therefore must be provided with the tools they need to successfully respond to gendered disinformation in real time, especially as it manifests itself in their election campaigns. Part of that defense involves understanding the online ecosystem where those harms proliferate and being at the fore-front of ongoing negotiations for new digital platform standards.

While some point to the importance of media literacy to build public immunity against misogyny and promote critical thinking, this does not represent a viable solution to gendered disinformation. According to cognitive scientist Stephan Lewandowsky: "This approach assumes that public misperceptions are due to a lack of knowledge and that the solution is more information – in science communication, it's known as the information deficit model. But that model is wrong: people don't process information as simply as a hard drive downloading data."[62] Misogynistic content, in particular, is designed to tap into emotionally loaded implicit bias against women in power. It's unlikely that fact-checking and media literacy will have much impact on altering this type of content, or its emotional effects on people. Due to the personal nature of gendered disinformation attacks, fact-checking often proves to be a challenge. Broader media literacy training requires widespread reception – yet even those trained in media literacy may not be immune to gendered disinformation. Most importantly, media literacy places the burden of the solution on individuals rather than on the platforms themselves, failing to address the root of the problem.

The International Foundation for Electoral Systems (IFES) recommends breaking the gender dimensions of disinformation into five components to find intervention points: actor, message, mode of dissemination, interpreters, and risk.[32] Civil society actors can play an important role in focusing in on the intervention points that are most relevant to the elections where attacks are organized, considering the cultural context of gendered disinformation in their countries.

Philanthropists have a key role to play too but must modernize their approach to democracy and technology and provide greater consideration of women's leadership. With increased philanthropic support to NGOs, civil society can play a critical role in organizing and informing women's "war rooms" to combat gendered disinformation, much like was organized voluntarily in the US 2020 election.

The Women's Disinformation Defense Project,[63] led by UltraViolet and comprised of membership-based gender justice organizations with real reach to society and media, used existing research and disinformation tracking infrastructure to develop systems for quickly spotting and responding to false narratives focused on gender, race, and the elections. They recognized that the first step to countering these narratives was identifying and understanding their evolution and growth, setting up social listening systems to identify nascent narratives and trends. They partnered with researchers to understand patterns in message distribution in order to develop the most effective strategies for countering gendered disinformation attacks. They provided clear, coordinated talking points and message content, creating a daily drumbeat of media talking points as well as a regular distribution of organic, shareable social media content to help drive a unified message. This content both responded to attacks, but more importantly advanced a positive message by lifting up women's status. Using public opinion message testing, they effectively empowered target voter groups to resist racist and misogynist messaging through offensive digital content. Holding the platforms responsible for their part in undermining the election and democracy was also a plank in their plan. They used people-powered mobilizations and earned media to broadly expose the harms caused by social media platforms to voters of color and women voters in order to try to change platform behavior in the short term, and laid the groundwork for a longer-term oversight.

Conclusion

As recent events in elections across the globe have made clear, gendered disinformation has gone mainstream and can easily intersect with violent extremism, with predictably tragic results. Gendered disinformation campaigns have become a primary strategy used by authoritarian regimes, emphasizing a concerning trend of the role of misogyny in the dismantling of democracy. Without increased attention and resources directed specifically toward combatting gendered disinformation, women's political involvement will continue to be blocked and democratic institutions will, in turn, continue to be undermined. By creating global frameworks to prevent gendered disinformation, establishing social media standards that remove current platform incentives, and providing women leaders the tools necessary for digital resilience, we can begin to ensure women's participation in leadership and the strength of our democratic systems. In absence of strong digital platform regulation, civil society and philanthropy must rise to the challenge and confront the threat that the undermining of women leaders poses to democratic values.

Notes

1 Maria Ressa, The Nobel Peace Prize 2021, Nobel Lecture, December 9, 2021, https://www.nobelprize.org/prizes/peace/2021/ressa/lecture/
2 Lucina Di Meco and Saskia Brechenmacher, "Tackling Online Abuse and Disinformation Targeting Women in Politics," November 30, 2020, accessed September 01,

2021, https://carnegieendowment.org/2020/11/30/tackling-online-abuse-and-disinformation-targeting-women-in-politics-pub-83331.

3 Sarah Repucci and Amy Slipowitz, *Freedom in the World 2021: Democracy Under Siege*, 2022 report, https://freedomhouse.org/sites/default/files/2022-02/FIW_2022_PDF_Booklet_Digital_Final_Web.pdf

4 Lucina Di Meco, "Online Threats to Women's Political Participation and The Need for a Multi-Stakeholder, Cohesive Approach to Address Them," (Paper presented at the Sixty-Fifth Session of the Commission on the Status of Women, October 5–8, 2020), 4.

5 Ibid.

6 Ellen Judson, Asli Atay, Alex Krasodomski-Jones, Rose Lasko-Skinner, and Josh Smith, *Engendering Hate: The Contours of State-Aligned Gendered Disinformation Online*, (London, UK: Demos, 2020), 7.

7 Declaration on the Elimination of Violence against Women Proclaimed by General Assembly resolution 48/104 of 20 December 1993

8 Dhanaraj Thakur and DeVan L. Hankerson, *Facts and Their Discontents: A Research Agenda for Online Disinformation, Race, and Gender*, (Washington, D.C.: Center for Democracy and Technology, 2021), 24.

9 Balmer, Crispian, Reuters, Top Italian Official Says Facebook Must Do More against Hate Speech, February 12, 2017

10 Ging (2017) as cited in Lisa Sugiura, *The Incel Rebellion: The Rise of the Manosphere and the Virtual War Against Women*, (U.K.: Howard House, 2021) 27, 28. https://library.oapen.org/handle/20.500.12657/51536

11 The definition of "abuse" is modeled from the Center for Democracy and Technology (CDT) and includes any direct or indirect threats of any kind, content that promotes violence against the individual based on any part of their identity (gender, race, ethnicity, religion, age), and content which attempts to demean and belittle the individual based on any part of their identity and that includes insults and slurs directed at the individual. Dhanaraj Thakur and DeVan L. Hankerson, *Facts and Their Discontents: A Research Agenda for Online Disinformation, Race, and Gender*, (Washington, D.C.: Center for Democracy and Technology, 2021), 24.

12 Jankowicz, Nina, Jillian Hunchak, Alexandra Pavliuc, Celia Davies, Shannon Pierson, and Zoë Kaufmann, *Malign Creativity: How Gender, Sex, and Lies are Weaponized Against Women Online*, (Washington, D.C.: Wilson Center, 2021).

13 *Barbara Lee Family Foundation, Keys to Elected Office: The Essential Guide for Women*, (Cambridge, MA: Barbara Lee Family Foundation, 2019).

14 Ibid.

15 Wegweiser – Hintergrundbericht für das Projekt »Radikalisierung in rechtsextremen Onlinesubkulturen entgegentreten, Dominik Hammer, Paula Matlach & Till Baaken, Veröffentlichung: 23. September 2021

16 Barbara Lee Family Foundation, *Keys to Elected Office: The Essential Guide for Women*, (Cambridge, MA: Barbara Lee Family Foundation, 2019).

17 "Catherine McKenna: Canada Environment Minister given Extra Security," BBC News, September 08, 2019, https://www.bbc.com/news/world-us-canada-49627153.

18 Ella Nilsen, ""Likability" Ratings in a Recent New Hampshire Poll Show Just How Tough Female Candidates Have It," Vox, July 23, 2019, https://www.vox.com/2019/7/23/20699724/likability-gender-new-hampshire-poll-warren-harris.

19 Ibid.

20 Alisha Haridasani Gupta, "The Likability Trap Is Still a Thing," *The New York Times*, November 22, 2019, https://www.nytimes.com/2019/11/22/us/the-likability-trap-women-politics.html.

21 Lean In, *How Outdated Notions about Gender and Leadership Are Shaping the 2020 Presidential Race*, (LeanIn.org, 2020).

22 Lucina Di Meco, *#ShePersisted Women, Politics, & Power in the New Media World: AI Two Pager*, (Washington, D.C.: The Wilson Center, 2019), 34.

23 Moya Bailey, *Misogynoir Transformed: Black Women's Digital Resistance*, (New York: NYU Press, 2021).

24 Candice Norwood, Chloe Jones, and Lizz Bolaji, "More Black Women Are Being Elected to Office. Few Feel Safe Once They Get There," PBS, June 17, 2021, https://www.pbs.org/newshour/politics/more-black-women-are-being-elected-to-office-few-feel-safe-once-they-get-there.
25 Abhishek Gupta. "The State of AI Ethics," Montreal AI Ethics Institute. January, 2021, 15–35.
26 Jankowicz et al., *Malign Creativity*, 1–55.
27 Maria Giovanna Sessa, "Misogyny and Misinformation: An Analysis of Gendered Disinformation Tactics During the COVID-19 Pandemic", Demos and EU Disinfo Lab, 2020, https://www.disinfo.eu/publications/misogyny-and-misinformation:-an-analysis-of-gendered-disinformation-tactics-during-the-covid-19-pandemic/
28 Lucina Di Meco, *#ShePersisted Women, Politics, & Power in the New Media World*, (Washington, D.C.: The Wilson Center, 2019), 1–58.
29 Flávia Biroli, "Violence against women in politics and public life, democratic backsliding, and far-right politics," (Paper presented at the Sixty-Fifth Session of the Commission on the Status of Women, September, 2020), 7.
30 Ellen Judson, "Gendered Disinformation: 6 Reasons Why Liberal Democracies Need to Respond to This Threat: Heinrich Böll Stiftung: Brussels Office – European Union," Heinrich-Böll-Stiftung, July 09, 2021, https://eu.boell.org/en/2021/07/09/gendered-disinformation-6-reasons-why-liberal-democracies-need-respond-threat?dimension1=democracy#GenderedDisinfo08.
31 Judson, Atay, Krasodomski-Jones, Lasko-Skinner, and Smith, *Engendering Hate: The Contours of State-Aligned Gendered Disinformation Online*, DEMOS National Democratic Institute, corp creators. (2020), https://dera.ioe.ac.uk/id/eprint/36786
32 Eliza Mackintosh, Swati Gupta, and Edward Scott-Clarke, "Troll Armies, 'Deepfake' Porn Videos and Violent Threats. How Twitter Became so Toxic for India's Women Politicians," CNN, January 23, 2020, https://edition.cnn.com/2020/01/22/india/india-women-politicians-trolling-amnesty-asequals-intl/index.html.
33 Barbara Lee Family Foundation, *MODERN FAMILY: How Women Candidates Can Talk About Politics, Parenting, and Their Personal Lives*, (Cambridge, MA: Barbara Lee Family Foundation, 2016).
34 Annika Neklason, "Moms Running for Office Are Finally Advertising Their Motherhood," *The Atlantic*, July 30, 2018, https://www.theatlantic.com/family/archive/2018/07/midterms-2018-mothers/565703/.
35 Kristina Wilfore and Lucina Di Meco, "Canadian Women Leaders' Digital Defense Initiative," Montreal Institute for Genocide and Human Rights Studies, June 3, 2021, https://www.concordia.ca/content/dam/artsci/research/migs/docs/Women-Leadership/WhitePaper_CanadianWomenLeaders.pdf
36 Astor, Maggie. "For Female Candidates, Harassment and Threats Come Every Day," *The New York Times*. August 24, 2018, https://www.nytimes.com/2018/08/24/us/politics/women-harassment-elections.html.
37 *Inter-Parliamentary Union, Sexism, Harassment, and Violence Against Women Parliamentarians*, (Geneva, Switzerland: Inter-Parliamentary Union, 2016).
38 Goldberg, Emma. "Fake Nudes and Real Threats: How Online Abuse Holds Back Women in Politics," *The New York Times*, June 4, 2021, https://www.nytimes.com/2021/06/03/us/disinformation-online-attacks-female-politicians.html.
39 Cécile Guerin and Eisha Maharasingam-Shah, *Public Figures, Public Rage: Candidate Abuse on Social Media*, (London, UK: Institute for Strategic Dialogue, 2020).
40 Laura Thornton, "Opinion | How Authoritarians Use Gender as a Weapon," *The Washington Post*, June 07, 2021, https://www.washingtonpost.com/opinions/2021/06/07/how-authoritarians-use-gender-weapon/.
41 Monika Bauhr, Nicholas Charron, and Lena Wängnerud, *Close the Political Gender Gap to Reduce Corruption*, (Norway: U4 Anti-Corruption Resource Centre, 2018). https://www.u4.no/publications/close-the-political-gender-gap-to-reduce-corruption.pdf.

42 Wagner, Angelia. "Tolerating the Trolls? Gendered Perceptions of Online Harassment of Politicians in Canada," Feminist Media Studies, April 8, 2020, https://www.tandfonline.com/doi/abs/10.1080/14680777.2020.1749691

43 Human Rights Council, *Report of the Working Group on the Issue of Discrimination against Women in Law and in Practice*, (Geneva, Switzerland: United Nations, 2018).

44 Rebecca Heilweil, "Why Algorithms Can Be Racist and Sexist," Vox, February 18, 2020, https://www.vox.com/recode/2020/2/18/21121286/algorithms-bias-discrimination-facial-recognition-transparency.

45 Frances Haugen testimony, United States Senate Committee on Commerce, Science and Transportation Sub-Committee on Consumer Protection, Product Safety, and Data Security, October 4, 2021, https://www.commerce.senate.gov/services/files/FC8A558E-824E-4914-BEDB-3A7B1190BD49

46 Keach Hagey and Jeff Horwitz, "Facebook Tried to Make Its Platform a Healthier Place. It Got Angrier Instead," Wall Street Journal, September 15, 2021, https://www.wsj.com/articles/facebook-algorithm-change-zuckerberg-11631654215

47 Daniel Funke and Susan Benkelman. "Factually: How Misinformation Makes Money," American Press Institute, September 26, 2019, https://www.americanpressinstitute.org/fact-checking-project/factually-newsletter/factually-how-misinformation-makes-money/.

48 Wallison, Peter. "Danger of Social Media Business Models," The Hill, https://thehill.com/opinion/technology/579369-danger-of-social-media-business-models

49 "The Twitter Rules: Safety, Privacy, Authenticity, and More," Twitter, help.twitter.com/en/rules-and-policies/twitter-rules.

50 Brandeis Marshall, "Algorithmic Misogynoir in Content Moderation Practice," Heinrich-Böll-Stiftung (June 2021).

51 "Facebook Community Standards," Transparency Center, Facebook, transparency.fb.com/policies/community-standards/?from=www.facebook.com/communitystandards.

52 Ysabel Gerrard and Helen Thornham, "Content Moderation: Social Media's Sexist Assemblages," *Sage Journals*, July 2020, 22 (7), 2020.

53 "Find Out Which Groups Get Big Tech Funding," Tech Transparency Project, August 25, 2021, https://www.techtransparencyproject.org/articles/find-out-which-groups-get-big-tech-funding.

54 Julia Carrie Wong, "How Facebook Let Fake Engagement Distort Global Politics: A Whistleblower's Account," *The Guardian*, April 12, 2021, https://www.theguardian.com/technology/2021/apr/12/facebook-fake-engagement-whistleblower-sophie-zhang.

55 Claire Cain Miller and Ruth Fremson, "They Believe in Ambitious Women. But They Also See the Costs," *The New York Times*, April 21, 2021, https://www.nytimes.com/interactive/2021/04/21/upshot/high-school-girls-politics.html?smid=tw-share.

56 *Re: UN Special Rapporteur's Annual Thematic Report to Be Presented to the Human Rights Council at Its 47th Session in June 2021*, Harvard Kennedy School Shorenstein Center to Irene Khan, February 15, 2021, https://documents-dds-ny.un.org/doc/UNDOC/GEN/G21/085/64/PDF/G2108564.pdf?OpenElement

57 ""Prioritise the Safety of Women": Open Letter to CEOs of Facebook, Google, TikTok & Twitter," Web Foundation to Mark Zuckerberg, Sundar Pichai, Shou Zi Chew, Jack Dorsey. July 1, 2021, https://webfoundation.org/2021/07/generation-equality-letter/

58 "Putting the Onus on Women is a PR Stunt—The Platforms Are the Problem," UltraViolet to Mark Zuckerberg, Sheryl Sandberg, Sundar Pichai, Susan Wojcicki, Jack Dorsey, Vanessa Pappas, July 1, 2021, https://weareultraviolet.org/putting-the-onus-on-women-is-a-pr-stunt-the-platforms-are-the-problem-2/

59 Eleanor Langford, "Politics Home," MP Damian Collins, July 29, 2020, https://damian-collins.com/politicshome-freedom-of-reach-is-not-the-same-as-freedom-of-speech/

60 "Online Harms White Paper: Full Government Response to the Consultation," (2020). https://www.gov.uk/government/consultations/online-harms-white-paper/outcome/online-harms- white-paper-full-government-response
61 ""Venerating the Housewife:" A Primer on Proud Boys' Misogyny," Anti-Defamation League, August 12, 2021, https://www.adl.org/blog/venerating-the-housewife-a-primer-on-proud-boys-misogyny.
62 Stephan Lewandowsky and John Cook, "The Debunking Handbook Part 1: The First Myth about Debunking," Shaping Tomorrows World, May 06, 2018, https://www.shapingtomorrowsworld.org/debunking-handbook-part-1-first-myth-about-debunking.html.
63 "Stop the Lies! WDDP to the Rescue Ongoing for Women in Politics," Philanthropy Women, November 2020, https://philanthropywomen.org/feminist-funding/stop-the-lies-wddp-to-the-rescue-ongoing-for-women-in-politics/.

PART IV

9

UNDERSERVED, UNDERREPRESENTED AND UNAWARE

Uplifting Women Through Digital Literacy Initiatives

Sun Sun Lim

Introduction

Techno-utopian discourse in the early days of the internet exulted in the emancipatory and democratising nature of the online milieu, lauding its ability to offer marginalised communities a viable platform for voicing their needs and a potent conduit for advancing their interests.[1] However, the passage of time has shown that the internet is hardly the much-vaunted liberating space where the oppressed can cast off the shackles of inequality. On the contrary, stratification and injustices that pervade the offline world are reproduced and even intensified online.[2][3]

Women, in particular, have found that chauvinism, sexism, misogyny and toxic masculinity have acquired a new vociferousness online.[4][5] Gender-based hate speech, online abuse and disinformation have proliferated, assuming novel forms ranging from the innocuous to the disturbingly graphic.[6] Emboldened by the cloak of anonymity, invigorated by the expanse of online networks and abetted by interactive multimedia capabilities, perpetrators of online harms against women have sunk to unchartered depths in intimidation of and violence against women.[7] The notorious Gamergate campaign of 2014 is a notable example, where male gamers targeted women in the video game industry and waged a coordinated online harassment campaign. Game developers Zoë Quinn and Brianna Wu, and feminist media critic Anita Sarkeesian, all of whom advocated for games to be less sexist and more progressive, bore the brunt of the harassment that included doxing and threats of rape, brutality and death.[8] Widespread sharing of explicit sexts, revenge porn images, videos of sexual brutality, online chats and memes denigrating females – these are just a small proportion of the online abuses women around the world encounter every day. Egregiously, some have even profited from enabling revenge porn such as the now defunct website isanyoneup.com that was created in 2010 by Hunter Moore and averaged

DOI: 10.4324/9781003261605-13

10,000 photo submissions over a three-month period while generating as much as USD13,000 in monthly advertising revenue.[9]

The rise in gender-based hate speech, online abuse and disinformation significantly constrains the online space that women can explore freely and autonomously, threatening their sense of security. This has adverse implications for women's full exploitation of the internet, thus hampering their educational progress, professional advancement, social networking and recreational pursuits. While it is thus evident that effort must be channelled into making the internet safer for women, what is less clear is which strategies and approaches will be most effective. In the face of online abuse, addressing the concerns and needs of women in the online realm is the definitive step towards designing a safer and more secure virtual environment.

However, the push to incorporate such needs into the design of online platforms will be ineffectual given the absence of any clear commercial incentives for big tech companies to do so. Tech company representatives may boldly proclaim their firm stances and swift action against hate speech being propagated on their platforms,[10] but industry insiders have gone on record to claim that the few ameliorative measures undertaken were superficial and ineffectual.[11] A more sustainable and enduring strategy is to shore up women's digital competencies to an extent that more women venture into the technology industry as entrepreneurs, scientists, designers, programmers and analysts. Ultimately, greater gender representation in the technology industry will translate not just into a sharpened sensitivity to women's needs, but concrete changes in the design and development of technological innovations. The next section will discuss in greater depth existing barriers to design and innovation that cater to gender differences.

As we work towards that longer-term goal of greater female representation in the technology sector, another clear and present imperative is to vest women with the knowledge of how they can help themselves and other women who suffer from online harms. To this end, advocacy for women by women can help raise awareness of online harms against women and possible remedies. Notably, in July 2021, more than 200 prominent women including Helen Clark, Thandiwe Newton and Gemma Chan signed an open letter calling for tech companies to "prioritize the safety of women" in conjunction with the Generation Equality Forum in Paris. Bringing together governments, businesses, international organisations and civil society to identify key priorities for advancing gender equality,[12] the forum subsequently saw Facebook, Google, TikTok and Twitter releasing commitments to improve the safety of women online.

By mobilising and empowering women to take more concerted action against online harms, we can better gauge the severity of the problem and accelerate the momentum to ameliorate, if not eradicate it. Hence, this chapter will delve into notable digital literacy programmes in Singapore targeted at females that seek to raise their digital literacy, be it for greater participation in technology-related work, or for improving awareness about avenues for redress when encountering online harms. The chapter will also assess these programmes so as to distil best practices with a view towards emulation by similar programmes in other countries. The three Singapore initiatives discussed in this chapter are: Daughters of Tomorrow: a skills and empowerment programme targeted at low-income women; Codette: a programming and digital skills training programme serving females in the minority Malay-Muslim community; and Solid Ground: a website offering step-by-step guides for victims of online harassment or abuse. The first two initiatives can help to boost national security because by equipping a broader swathe of the underserved population with digital competencies, they will be less susceptible to scams, misinformation and disinformation that is deliberately designed by bad actors to manipulate and mislead audiences. The last initiative will vest women with the knowledge of how to seek redress if experiencing online harms and such insights can help to augment their individual security. In totality, all these efforts at enhancing women's digital literacy also boost national security because they empower women and, in turn, benefit the next generation of Singaporeans in view of the vital role women play in nurturing our youth. Overall, Singapore also provides an interesting context for investigation because of the country's avid adoption of technology, strong rule of law and its efforts in promoting gender equality.[13]

How women in technology can make a difference

The genesis and development of popular dating app Bumble – with over 100 million subscribers worldwide as of 2020 – illuminates why it is vital for women to gain a stronger foothold in the technology sector, and the difference this can make. Founder and CEO Whitney Wolfe Herd built on her experience of working at rival dating app Tinder to create Bumble, a female-focused dating app where only women can initiate the first contact with a prospective date.[14] This seemingly nondescript feature is tremendously empowering because it helps women avoid being bombarded with sexually-charged and demeaning messages from random users and puts them in control of the relationship-building process. In so doing, it also helps to redefine how men and women approach dating both online and offline.

Bumble's female-friendly features and its overarching mission to make dating safer for women were clearly grounded in Herd's experience as an eligible young woman seeking meaningful relationships. She had endured an abusive relationship as a teenager that gave her a deep insight into problems arising from gender dynamics. Her success demonstrates how catering to women's needs and interests can serve diverse consumer groups and contribute to the company's growth, while also setting new industry benchmarks. Notably, in November 2020, Bumble introduced a feature to prevent bad actors from using the app's "unmatch" feature to hide from victims, leading Tinder to introduce an analogous feature in response.[15] Herd's journey shows that with more women at the helm to ensure that women's views, insights, and interests are represented and met by technological innovations, more inclusive and secure technology is achievable.

It is thus evident that greater gender representation in the technology industry will not just translate into sharpened sensitivity to women's needs and wants, but concrete changes in the design and development of technological innovations. The technology we use today is heavily biased towards male needs and perspectives. There is extensive evidence of the gender-biased nature of technology design and production, and the male-dominated environment in technology hubs such as Silicon Valley overlooking the needs and views of females, sometimes with serious consequences.[16] Crash test dummies that are based on the average male physique, ill-fitting armour that fails to protect policewomen and cancer research based principally on data on the male immune system – these are all sobering examples of how the omission of women's needs can cost lives.[17] In terms of national security, the disproportionate burden that women bear as caregivers to children and the elderly means that they play a pivotal role in supporting and guiding the digital practices of these vulnerable groups. Hence, women must be mobilised to help raise the community's overall defences against cyberscams, online terrorism, ideological extremism and online disinformation.

While the case for a greater female representation in tech is an inherently compelling one, the path towards this goal is far from rosy. In the first instance, unless we actively shore up women's digital competencies, they will feel less prepared for, and thus less drawn to employment opportunities in the technology sector.[18] Worldwide, women's participation in Science, Technology, Engineering and Mathematics (STEM) education and in the technology sector workforce is systematically lower than in other industries. Of technology majors in Southeast Asia for example, 39% are women (compared with 56% for all other fields of study).[19] Correspondingly, in the region's workforce, women constitute 38% of the total workforce but only 32% of the technology sector.[20] Singapore is above the regional average with about 41% of women working in tech although it has one of the lowest shares of women majoring in tech.[21] In terms of technology entrepreneurship, women have even more catching up to do. Worldwide, only 20% of all technology start-ups were founded by women and as of 2014, only 17.4% of CIOs in Fortune 500 companies were women.[22] In light of these modest numbers, we need robust and creative solutions to augment and deepen women's digital competencies so that they can become equal players in the technology industry as entrepreneurs, scientists, programmers, designers, analysts and more.

Even as attracting women to join the tech sector is a challenge, retaining them in the industry is another monumental hurdle and presents a pipeline issue.[23] While the number of women entering the technology sector is not sizeable to begin with, attrition rises sharply after pregnancy and childcare. In a fast-moving sector such as technology, the task of reintegrating is significantly more daunting as one's skills become quickly outdated. Furthermore, because there are fewer women to rise to the top, younger women do not see sufficient female role models in the technology sector that is not very female-friendly to begin with. They are thus discouraged from joining the sector or leave prematurely, uncertain of their prospects.

Upskilling policies such as Singapore's SkillsFuture credit programme can certainly help women to upskill for growth sectors and industries. However, better support is also needed for women who take leave for pregnancy, childcare or eldercare to help them sustain and broaden their professional networks. These will smoothen their path as they return to the workforce. We should explore public and private sector support for such women's membership in professional societies. States should also consider providing special incentives such as tax breaks for companies in sectors where women are under-represented to motivate the introduction of more flexible working arrangements and better childcare support. We should also foster more incubation schemes that take into account the circumstances of female innovators, and that offer more systematic mentoring so that more women engage in entrepreneurship in the tech sector.

More importantly, states and academic institutions should engage in systematic longitudinal research that tracks the career trajectories of women in the technology sector so as to better understand their reasons for attrition, as well as factors that promote re-entry into the workforce. It is only with robust evidence that we can take more proactive measures to ameliorate this problem of female under-representation in tech.

The relationship between women in tech and workplace harassment, be it offline or online, also warrants a closer scrutiny. Women in technology companies like e-commerce giant Alibaba and popular game developer Activision Blizzard have reported allegations of workplace and sexual harassment by male colleagues in positions of seniority.[24] According to the organisation Women Who Tech, 48% of women working in the technology sector experienced harassment, of whom 43% said they were sexually harassed. Worryingly, women in tech appear to be growing more reticent about reporting harassment, with 45% reporting in 2020 versus 55% in 2017.[25] In Singapore, according to a study by Ipsos and the Association of Women for Action and Research (AWARE),[26] only one in three victims (both women and men polled) reported harassment to their boss, a senior person at work or the HR department. While there are no official sexual harassment statistics on women in Singapore's tech sector, harassment purportedly goes unreported because victims are deterred by potential damage to career advancement. A female Singapore-based director of Entrepreneur First (a platform that builds deep-tech companies) was quoted as saying, "The incentive structure for flagging harassment is wrong: as a victim you're expected to speak up, and most successful women in tech I know haven't worked this hard and defied all the odds just to have their reputation tarnished over a sexual discrimination case. The repercussions are real".[27]

In light of our digitally-connected workplaces, it is common for workplace harassment to take place via WhatsApp, office messaging services and social media.[28] These unwelcome advances can be as innocuous as male superiors demanding that female employees appear on camera during Zoom meetings when it is not required so that they can ogle at their subordinates, to stalking a colleague by referring to their calendars[29] on email applications like Microsoft Outlook. Chat technologies or online communication services such as WhatsApp, Slack and Skype can also be used as platforms for workplace harassment as they transcend work-life boundaries.[30] Despite work from home becoming more prevalent due to the pandemic, incidences of workplace sexual harassment have not declined with the reduction in physical interaction. One study found that only 17% of respondents polled in Singapore reported a decrease in incidents of offline sexual harassment while 7% reported an increase in online sexual harassment.[31] The migration of workplace sexual harassment to the online realm introduces novel challenges. The absence of bystanders and witnesses in an online work environment may further embolden perpetrators[32] as they believe they can get away with it. All companies need to be more proactive about setting codes of conduct for online communication among colleagues, and to take a zero-tolerance stance against workplace harassment whether online or offline. With technology companies in particular being so reliant on online communication, they should also leverage their technological prowess to create safer communication platforms equipped with reporting functions that assure harassment victims and whistle-blowers of anonymity.

Indeed, technology companies have simply not responded more robustly in addressing the technology-facilitated violence, which includes contact-based harassment, image-based abuse and gender-based cyberhate.[33] In Singapore, news emerged of Telegram chat groups such as "SG Nasi Lemak" where members circulated non-consensual and voyeuristic pictures of women and girls taken without their knowledge.[34] [35] Such chats can serve to normalise misogynistic behaviour and shrink the online space for women. Indeed, online violence is one of the leading causes of the gender digital divide globally.[36] [37] Even online threats may discourage women, especially younger girls, from accessing the internet.[38] This may lead to the occurrence of technophobia and anxiety towards technology, that coupled with existing gender stereotypes,[39] [40] may discourage females from venturing into STEM fields.[41] Fundamentally, if the technology sector does not take resounding action against workplace harassment, women will be deterred from considering careers in technology, further denting diversity and inclusion efforts.

Boosting women's digital competencies: Key Singapore initiatives

As an intensely technologising country with an active female workforce,[42] Singapore has seen the emergence of several digital literacy initiatives that support females. This chapter will highlight three notable examples. One offers skills and empowerment programmes to low-income women so as to enhance their employability and social mobility, another provides minority women with training and mentorship so that they can make inroads into the technology sector, and the last dispenses a comprehensive range of information on support and remedial action to guide women who have encountered online harassment.

Aiding the underserved: Daughters of tomorrow

Established with the goal of empowering underprivileged women, Daughters of Tomorrow (DoT) aims to help them boost their financial independence and nurture resilient families. Its beneficiaries are women aged 20–60 from low-income families averaging $200–$500 per capita per month and who mostly live in government subsidised rental apartments.[43] Many face multiple stressors in their lives, with difficulty in seeking gainful, stable employment being the key challenge. Besides their modest skillsets and rigid employment practices in general, many of these women shoulder significant childcare responsibilities that further limit the range of jobs available to them.[44]

In a rapidly digitalising economy like Singapore's, an impressive range of IT literacy training programmes are on offer in formal education, as well as through national skills upgrading programmes such as SkillsFuture. However, such programmes are pitched above the level of these low-income women who lack previous IT exposure and job experience but who require training in foundational skills.[45] DoT has thus sought to plug this critical gap by providing these underserved women with infocomm skills via its IT Literacy Programme's basic and advanced IT courses.[46] Trainees are taught to set up an email account and are introduced to the fundamentals of the internet, before being schooled in essential and popular tools such as Microsoft Word and Excel. Public and corporate donations are also tapped to provide the women with second-hand laptops and software subscriptions so that they can practise their IT skills at home.[47] By vesting them with these computer skills and devices, they are better qualified to apply for a wider selection of jobs in a digitalising environment. Furthermore, clerical and administrative jobs that demand such skills are also easier to perform from home, thereby making it easier for these women to juggle their work and family commitments. It is worth noting that DoT factors in these women's familial obligations and makes special arrangements for childcare during the training sessions so that trainees can be fully committed to learning.[48]

While the espoused goals of DoT's IT literacy programmes are to help these women improve their employability, one very distinct subsidiary benefit is these women's improved capacity to guide their children's IT use and to therefore become more effective parents in this important regard. Digitalisation has swept through education and schools are more avidly using online learning platforms to teach and assess children, as well as to communicate with parents about children's progress.[49] Infocomm-illiterate parents are particularly disadvantaged because they lack the skills to support their children's learning, or to guide their children's optimal exploration of the internet for entertainment and social interaction.[50] With more infocomm-literate parents in the low-income group, it will be easier to shore up the IT literacy of underprivileged children and further safeguard them against online harms such as harassment, scams, falsehoods and recruitment by extremist organisations. Fundamentally, since the home is the child's principal environment for media consumption, parents (especially mothers) play a critical role in guiding children's digital literacy skills but only if they possess the technical competencies to begin with.[51] Empowering these low-income women with digital skills will in turn empower their children as well.

Uplifting the underrepresented: The Codette Project

The Codette Project was founded in 2015 as a non-profit organisation with seed-funding from Mendaki, a Malay-Muslim self-help group that supports students and individuals with education and training to uplift the community's educational performance. Targeted at minority Muslim women, Codette's founder Nurul Jihadah recognised that her community was severely underrepresented in the tech industry. She lamented that in Singapore, "fewer than 5 per cent of start-up founders are women, and some minorities, like the Malay-Muslim minority, are practically invisible".[52] She sought inspiration from NGOs outside of Singapore and was struck by the experience of Black Girls Code, a movement in the United States for young black girls to nurture their interest and competencies in technology and programming.[53] Indeed, given that gender inclusion policies and corporate practices can vary widely across countries, the sharing and showcasing of efforts that successfully uplift ethnic minority women from across the globe can help to ignite and advance local efforts.[54]

Codette organises a diverse range of courses, discussion panels and workshops for minority women of all backgrounds to share how technology can improve their lives. They focus on topics such as web design, social media branding, HTML, data analytics and user experience design. For example, Instagram Story School helps small business owners learn promotional strategies for reaching out to consumers on the visually-rich platform. Notably, Codette also hosts mentoring events so that women who have succeeded in technology can impart their know-how and inspire other women to follow in their footsteps. Through this diverse range of efforts, Codette has played a pivotal role in expanding the ecosystem of minority women with the skills, passion and networks to mentor and inspire others to join the technology sector.

This concerted cultivation of underrepresented minority women is critical for technology design that takes into account the full spectrum of diverse voices. If minority women are left out of the equation, the silencing of their experiences and insights in future innovation will be detrimental to the furtherance of inclusive design. The limited research on ethnic minority women in STEM that adopts intersectional approaches suggests that they encounter a "double jeopardy" in having to fight negative stereotypes about their competence based on both their gender and their race.[55] While such STEM-focused research has yet to be conducted in the Singapore context, other studies have found that

Singapore's ethnic minorities do face or perceive workplace racial discrimination when it comes to seeking employment or scoring a promotion.[56] It is not unlikely for such discrimination to be perceived in Singapore's technology sector as well. When any segment of the population feels marginalised, the likelihood of simmering resentment and social fissures being exploited by bad actors to foment discord is raised. The proactive inclusion of ethnic minority women in the technology sector, and a boost to their digital competencies can therefore help to reinforce national security.

Supporting the unaware: Solid ground

Solid Ground is a website that provides helpful step-by-step guides for victims of online harassment or abuse in Singapore. These offer detailed instructions for, *inter alia*, ensuring one's personal safety, gathering incriminating evidence, blocking and reporting perpetrators, obtaining support and legal assistance, and making a police report. The guides cover nine typical online harassment situations such as when one's intimate images have been taken and/or shared without permission, one's personal information has been divulged online or when one is being repeatedly contacted or stalked. A separate section provides information on government agencies, voluntary welfare organisations, charities or educational institutions from which victims can seek help. This information is classified according to the different communities these organisations serve – women, LGBTQ, men and university students.

These resources were collated and developed by two researchers at the Singapore University of Technology and Design's Lee Kuan Yew Centre for Innovative Cities, Catherine Chang and Holly Apsley. The guides were developed and informed by their interviews with individuals who had experienced online harms and social workers supporting such victims. The two founders run Solid Ground as a voluntary, independent project with the support of Singapore's leading advocacy group for women's rights, AWARE, and the National Youth Council.

Chang and Apsley were motivated to start Solid Ground when they realised that Singapore lacked online resources to support victims of online harms, even though these were widely available and well-established in other countries. They share the conviction that making such information widely available will raise awareness that such harassment is unacceptable and prevent it from being normalised. A key challenge they faced was striking the right balance between the website being comprehensive but not overwhelming, yet offering actionable guidance. Since its launch, they have received very positive feedback and shows of support such as Solid Ground being promoted by reputable educational Instagram accounts as a trustworthy resource. AWARE has also been recommending Solid Ground to its clients. The founders plan to further promote the site's content in even more accessible formats such as via social media posts, and to work with support helplines in Singapore to enhance their FAQs for people seeking advice on online harms.

Best practices and recommendations

These three examples of notable digital literacy initiatives in Singapore that empower women offer invaluable learning points for similar endeavours in other parts of the world. They share various strengths, the first of which is having clearly identified target groups and stakeholders. Rather than serving generic and diffuse populations, they have zeroed in on female communities with distinct difficulties. Such concentrated attention enables them to be more deeply engaged with their clients and their particular circumstances. Hence, a second notable strength is that they have assiduously sought to understand the needs and concerns of the groups they serve. Notably, the empathy that their founders have for the women they empower, and the challenging environments these women inhabit, ensure that their digital literacy efforts especially resonate. DoT's proactive provision of childcare for their trainees' children, Codette's focus on networking and mentoring for Malay-Muslim women who lack such role models and Solid Ground's packaging of their guides into scenarios of online harms women typically face – these examples demonstrate that when women are in the driver's seat, they bring their deep insights to bear on the organisations they lead, the programmes they launch and the communities they support. Third, all three initiatives have sharply focused and clearly articulated visions that have provided a lodestone for their ongoing efforts and future plans. By building on their earlier accomplishments while focusing on their overarching vision, they can become progressively more ambitious, yet targeted. The experiences of these three programmes, while instructive, will most likely be successfully replicated in countries where women's civil rights are a given, and where access to education and digital connectivity are forthcoming.

Their experiences also suggest that there are many purposeful and strategic ways to raise women's digital literacy and boost their participation in the technology sector. It is heartening that these initiatives have received support from both private and public sector organisations. Given the discernible societal benefits that such programmes can bring about, we need to consider other ways in which groups and individuals with similar visions can be supported policy-wise. The government can play an important stewardship role in bringing together relevant stakeholders and providing resources for more systematic collaboration. For example, Singapore's Ministry of Communication and Information engaged more than 300 individuals on issues like technology-facilitated gender-based harassment and online platforms that encourage vice and harm.[57] Following which it launched an Alliance for Action – people, private and public sector coalitions working in partnership – to tackle online harms against women and girls.

Another important responsibility of the state is to provide the necessary resources for complementary initiatives. One possibility is for relevant government agencies to offer grants that are specifically ring-fenced for digital literacy programmes focusing on women and other marginalised communities. Another avenue is to explore collaboration and data sharing so that these organisations can be better informed as to the communities they serve and can fine-tune their services to a higher degree. For example, the Ministry of Manpower and its associated jobs and skills agencies, Workforce Singapore and SkillsFuture Singapore respectively, can draw on their data repositories to identify specific technology sectors in which women are under-represented and share such insights with entities such as DoT and Codette so that they can adjust their training accordingly. Tax breaks and other corporate incentives should also be offered to companies that introduce gender-inclusive hiring practices and invest in creating gender-inclusive products and services.

Only by experimenting with tactics and solutions on multiple fronts can we more convincingly move the needle on greater female representation in the technology sector. And only then can we have technology products and services that reflect the needs of all and not just half of humanity. Societally, we must also amplify the discourse around the virtues of catering to women's needs and interests in technology design. Besides leveraging moral suasion and emphasising that gender inclusion is the right thing to do, we must further underline the fact that gender-inclusive companies will heighten their reputations for corporate social responsibility and this can translate into commercial gain as well.

The endeavour to broaden the participation of women in the technology sector will therefore pay dividends across all sectors of our society. Ultimately, a more secure and welcoming online milieu for women will also make the internet safer for all users.

Notes

1 Shirley Turkle, "Life on the Screen: Identity in the Age of the Internet", 1995, New York: Simon & Schuster.
2 Emily Harmer and Karen Lumsden, "Online Othering: An Introduction", in *Online Othering*, 2019, pp. 1–3, Cham: Palgrave Macmillan.
3 Jacqueline Ryan Vickery, "This Isn't New: Gender, Publics, and the Internet", in *Mediating Misogyny: Gender, Technology, and Harassment*, 2018, pp. 37–39, Cham: Palgrave Macmillan.
4 Ruth Lewis, Michael Rowe and Clare Wiper, "Online Abuse of Feminists as an Emerging Form of Violence Against Women and Girls", *British Journal of Criminology* 57, no. 6, 2017, pp. 1462–1481, https://doi.org/10.1093/bjc/azw073
5 Nithya Sambasivan, Amna Batool, Nova Ahmed, Tara Matthews, Kurt Thomas, Laura Sanely Gaytán-Lugo, David Nemer, Elie Bursztein, Elizabeth Churchill and Sunny Consolvo, "They Don't Leave Us Alone Anywhere We Go - Gender and Digital Abuse in South Asia", in *Proceedings of the 2019 CHI Conference on Human Factors in Computing Systems*, 2019, pp. 1–14.
6 George Veletsianos, Shandell Houlden, Jaigris Hodson and Chandell Gosse, "Women Scholars' Experiences with Online Harassment and Abuse: Self-Protection, Resistance, Acceptance, and Self-Blame", *New Media & Society* 20, no. 12, 2018, p. 4689, https://doi.org/10.1177/1461444818781324.
7 Jacqueline Ryan Vickery and Tracy Everbach, "The Persistence of Misogyny: From the Streets, to our Screens, to the White House", in *Mediating Misogyny: Gender, Technology, and Harassment*, 2018, pp. 1–27, Cham: Palgrave Macmillan.
8 Kishonna L. Gray, Bertan Buyukozturk and Zachary G. Hill, "Blurring the Boundaries: Using Gamergate to Examine 'Real' and Symbolic Violence against Women

in Contemporary Gaming Culture", *Sociology Compass* 11, no. 3, 2017 e12458, p. 22, https://doi.org/10.1111/soc4.12458

9 Samantha Bates, "Revenge Porn and Mental Health: A Qualitative Analysis of the Mental Health Effects of Revenge Porn on Female Survivors", *Feminist Criminology* 12, no. 1, 2017, pp. 22–42. https://doi.org/10.1177/1557085116654565

10 Nick Clegg, "Facebook Does Not Benefit From Hate", *Facebook*, 1 July 2020, https://about.fb.com/news/2020/07/facebook-does-not-benefit-from-hate/

11 Karen Hao, "She Risked Everything to Expose Facebook. Now She's Telling Her Story", *MIT Technology Review*, 29 July 2021, https://www.technologyreview.com/2021/07/29/1030260/facebook-whistleblower-sophie-zhang-global-political-manipulation/

12 Kirsten Salyer, "200 Women Call on Tech Giants to Prioritize Online Safety. Here's how", *World Economic Forum*, 2 July 2021, https://www.weforum.org/agenda/2021/07/200-women-call-on-tech-giants-to-prioritize-safety-online-heres-how/

13 Theresa W. Devasahayam (Ed.), "Singapore Women's Charter: Roles, Responsibilities, and Rights in Marriage", 2011, Singapore: Institute of Southeast Asian Studies.

14 Avery Hartmans and Annabelle Williams, "How Bumble Grew From a Female-Focused Dating App to a Global Behemoth Valued at Over $8 Billion After Going Public", *Business Insider,* 12 Feb. 2021, https://www.businessinsider.com/bumble-dating-app-company-history-2021-ipo-2020-9

15 Sarah Perez, "Bumble's New Feature Prevents Bad Actors From Using 'Unmatch' to Hide From Their Victims", *TechCrunch*, 10 November 2020, https://techcrunch.com/2020/11/09/bumbles-new-feature-prevents-bad-actors-from-using-unmatch-to-hide-from-their-victims/

16 Caroline Criado Perez, "Invisible Women: Exposing Data Bias in a World Designed for Men", 2019, London: Penguin Random House.

17 Ibid.

18 Karen E. Mishra, Kelly Wilder and Aneil K. Mishra. "Digital Literacy in the Marketing Curriculum: Are Female College Students Prepared for Digital Jobs?" *Industry and Higher Education* 31, no. 3, 2017, pp. 204–211, https://doi.org/10.1177/0950422217697838

19 Statista, "Full-Time Employment in the Information and Communication Technology (ICT) Industry Worldwide in 2019, 2020 and 2023", Statista, 17 June 2021, https://www.statista.com/statistics/1126677/it-employment-worldwide/

20 Ibid.

21 Olivia Poh, "Number of Women in Tech in South-East Asia Beats Global Average", *Business Times*, 20 October 2020, https://www.businesstimes.com.sg/garage/news/number-of-women-in-tech-in-south-east-asia-beats-global-average

22 Sophie Deering, "The Role of Women in the Tech Industry Today", Undercover Recruiter, 17 June 2021, https://theundercoverrecruiter.com/role-women-tech/

23 Erin Carson, "Half of Young Women will Leave Their Tech Job by Age 35, Study Finds", *CNet*, 29 September 2020, https://www.cnet.com/news/half-of-young-women-will-leave-their-tech-job-by-age-35-study-finds/

24 Jamilah Lim, "What's with Tech Companies and Workplace Harassment?", *Tech Wire Asia*, 17 August 2021, https://techwireasia.com/2021/08/whats-with-tech-companies-and-workplace-harassment/

25 Women Who Tech, "The State of Women in Tech and Startups", Women Who Tech, 6 January 2022, https://womenwhotech.org/data-and-resources/state-women-tech-and-startups

26 Theresa Tan, "Two in Five Workers in Singapore Say They Have Been Sexually Harassed at Work: Aware Study", *The Straits Times*, 15 January 2021, https://www.straitstimes.com/singapore/two-in-five-workers-said-they-have-been-sexually-harassed-at-work-aware-study

27 Jacquelyn Cheok, "Tech: No (Wo) Man's Land?", *The Business Times*, 9 September 2017, https://www.businesstimes.com.sg/brunch/tech-no-wo-mans-land

28 "Harassment at the Workplace - What Women in Singapore Need To Know", Girls in Tech Singapore, 6 January 2022, https://singapore.girlsintech.org/harassment-at-the-workplace-what-women-in-singapore-need-to-know/

29 Kathy Gurchiek, "Take Precautions Against Online Harassment in Virtual Workplaces", *SHRM*, 12 March 2021, https://www.shrm.org/resourcesandtools/hr-topics/technology/pages/take-precautions-against-online-harassment-in-virtual-workplaces.aspx

30 Nelson Tenório and Pernille Bjørn, "Online Harassment in the Workplace: The Role of Technology in Labour Law Disputes", *Computer Supported Cooperative Work (CSCW)* 28, no. 3, 2019, pp. 293–315. https://doi.org/10.1007/s10606-019-09351-2

31 Stephen Tracy, "Should Companies Rethink Their Workplace Harassment Policies with a Shift to Hybrid or WFH Arrangements?", *Milieu Insight*, 4 October 2021, https://mili.eu/insights/should-companies-rethink-their-workplace-harassment-policies

32 Leah Fessler, "Workplace Harassment in the Age of Remote Work", *The New York Times*, 8 June 2021, https://www.nytimes.com/2021/06/08/us/workplace-harassment-remote-work.html

33 Laura Vitis, "Technology-Facilitated Violence Against Women in Singapore: Key Considerations", in Jane Bailey, Asher Flynn and Nicola Henry (Eds.) *The Emerald International Handbook of Technology-Facilitated Violence and Abuse*, 2021, Bradford: Emerald Publishing Limited, p. 407.

34 Prisca Ang, "Police Looking into Another Telegram Chat Group Allegedly Circulating Obscene Materials", *The Straits Times*, 20 October 2019, https://www.straitstimes.com/singapore/police-looking-into-another-telegram-chat-group-allegedly-circulating-obscene-materials

35 Laura Vitis, "Technology-Facilitated Violence Against Women in Singapore: Key Considerations", in Jane Bailey, Asher Flynn and Nicola Henry (Eds.) The Emerald International Handbook of Technology-Facilitated Violence and Abuse, 2021, Bradford: Emerald Publishing Limited, p. 411.

36 Hyeonsoo Jeon, "Cyberviolence Disempowers Women and Girls and Threatens Their Fundamental Rights", *United Nations Development Programme*, 25 November 2021, https://www.eurasia.undp.org/content/rbec/en/home/blog/2021/Cyberviolence.html

37 Konstantina Davaki, *The Underlying Causes of the Digital Gender Gap and Possible Solutions for Enhanced Digital Inclusion of Women and Girls*, 2018, European Union: Policy Department for Citizens' Rights and Constitutional Affairs, Luxembourg: p. 22. 6 January 2022, https://www.europarl.europa.eu/RegData/etudes/STUD/2018/604940/IPOL_STU(2018)604940_EN.pdf

38 Ibid, 36.

39 Ibid.

40 José Luis Martínez-Cantos, "Digital Skills Gaps: A Pending Subject for Gender Digital Inclusion in the European Union", *European Journal of Communication* 32, no. 5, 2017, p. 435, https://doi.org/10.1177/0267323117718464

41 Hyeonsoo Jeon, "Cyberviolence Disempowers Women and Girls and Threatens Their Fundamental Rights", *United Nations Development Programme*, 25 November 2021, https://www.eurasia.undp.org/content/rbec/en/home/blog/2021/Cyberviolence.html

42 Ministry of Social and Family Development, "Labour Force and the Economy: Labour Force Participation Rate", 27 July 2021, https://www.msf.gov.sg/research-and-data/Research-and-Statistics/Pages/Labour-Force-and-the-Economy-Labour-Force-Participation-Rate.aspx

43 Lianne Ong, "Building a Kampung of Trust: Women Helping Women with a Win-Win Childcare Solution", *The Pride*, 22 August 2020, https://pride.kindness.sg/a-win-win-childcare-solution/

44 Callum Laing, "Carrie Tan, Founder of Daughters of Tomorrow (DOT)", *Empirics Asia*, 5 August 2015, https://empirics.asia/carrie-tan-founder-of-daughters-of-tomorrow-dot/

45 Daughters of Tomorrow, "IT Literacy", Daughters of Tomorrow, 27 July 2021, https://daughtersoftomorrow.org/get-support/it-literacy-program/

46 Ibid.

47 Xing Hui Kok, "Women from Poorer Families Pick Up IT Skills", *The Straits Times*, 5 September 2016, https://www.straitstimes.com/singapore/manpower/women-from-poorer-families-pick-up-it-skills

48 Lianne Ong, "Building a Kampung of Trust: Women Helping Women with a Win-Win Childcare Solution", *The Pride*, 22 August 2020, https://pride.kindness.sg/a-win-win-childcare-solution/

49 Sun Sun Lim, *Transcendent Parenting: Raising Children in the Digital Age*. 2020, New York: Oxford University Press.

50 Sun Sun Lim and Renae Loh Sze Ming, "Young People, Smartphones, Invisible Illiteracies: Closing the Potentiality-Actuality Chasm in Mobile Media", in Erika Polson, Lynn Schofield Clark and Radhika Gajjala (Eds.), *The Routledge Companion to Media and Class*, 2020, pp. 132–141. New York: Routledge.

51 Sun Sun Lim, "Through the Tablet Glass: Mobile Media, Cloud Computing and Transcendent Parenting", in Dafna Lemish, Amy Jordan and Vicky Rideout (Eds.), *Children, Adolescents, and Media - The Future of Research and Action*, 2017, pp. 18–26. London: Routledge.

52 Sue Ann Tan, "Causes Week 2017: Code for Success: Getting Women to Be Techies", *The Straits Times*, 10 December 2017. https://www.straitstimes.com/singapore/code-for-success-getting-women-to-be-techies

53 Nurul Hazirah, "They Are Paving the Way for Minority Women to Be Successful Through Tech", *YP SG*, 20 December 2018, https://www.yp.sg/codette-project-tech-sg/

54 Jie Shen, Ashok Chanda, Brian D'netto and Manjit Monga, "Managing Diversity Through Human Resource Management: An International Perspective and Conceptual Framework", *The International Journal of Human Resource Management* 20, no. 2, 2009, pp. 235–251. https://doi.org/10.1080/09585190802670516

55 O'Brien, Laurie T., Donna M. Garcia, Alison Blodorn, Glenn Adams, Elliott Hammer and Claire Gravelin, "An Educational Intervention to Improve Women's Academic STEM Outcomes: Divergent Effects on Well Represented vs. Underrepresented Minority Women", *Cultural Diversity and Ethnic Minority Psychology* 26, no. 2, 2020, pp. 163–168. https://doi.org/10.1037/cdp0000289

56 Matthew Mathews, "Indicators of Racial and Religious Harmony: An IPS-OnePeople.sg Study", Institute of Policy Studies, *Singapore: Institute of Policy Studies and Lee Kuan Yew School of Public Policy, National University of Singapore*, 2 August 2013.

57 Hwee Min Ang, "New Alliance for Action to Tackle Online Harms, Especially Those Targeted at Women and Girls", *ChannelNewsAsia*, 21 July 2021, https://www.channelnewsasia.com/singapore/online-sexual-harassment-alliance-for-action-for-girls-2044811

10

A NEW ERA IN THE FIGHT AGAINST ONLINE MISOGYNY

Priyank Mathur

My work over the past five years has taken me on a series of adventures – from Bollywood film sets in Mumbai to the dense forests of Mindanao, Philippines, from writing scripts with female comedians in Jakarta to running digital literacy workshops for girls in Davao City. Through it all, my colleagues and I at Mythos Labs have been honoured to partner with global institutions such as UN Women, the US Department of State and the European Commission in order to develop new approaches towards combatting very old problems – gender inequality and misogyny. Mythos Labs is a social enterprise I founded in 2017. It uses media and technology to counter gender inequality, violent extremism and mis/disinformation around the world. In this chapter, I will first share the findings of a study that Mythos Labs conducted with the support of UN Women to understand the extent of the threat of online misogyny and hate speech against women in Asia, especially in the context of COVID-19. Then, I will explain how my organization is using humour and digital literacy to design effective counter-measures.

Researching the continued threat of online misogyny and hate speech against women

In the Summer of 2020, as the pandemic's first wave led governments to declare national lockdowns and limit social interaction for billions of people, social media usage skyrocketed. Monthly average users for the world's most popular social media sites increased by over 21% from 2019 to 2021.[1] We at Mythos Labs decided to investigate what effect, if any, the COVID-19 pandemic and increased social media usage was having on gender-based hate speech and online misogyny. With support from UN Women Asia, our team of researchers analysed potential linkages between COVID-19 and online hate

DOI: 10.4324/9781003261605-14

speech against women in five countries with high numbers of social media users – India, Indonesia, Malaysia, Sri Lanka and the Philippines.[2] We used our MIDAC technology and qualitative research to study Twitter, Facebook and Google Search data. Using opinion dynamics modelling and machine learning, we also measured how much misogynist accounts were influencing the opinions of their Twitter audience.

Our research revealed three key findings.[3] Firstly, the volume of misogynistic posts, tweets and Google searches was significantly higher during national lockdowns (March – June of 2020), as compared to the same time period in 2019. Secondly, we found that across India, Sri Lanka and Malaysia, sexist organizations, i.e. organizations committed to espousing anti-women rhetoric, were disseminating new, COVID-related misogynist narratives on Facebook. Some of the most popular narratives included "COVID is exposing the ugly truth about women", "women are more dangerous than men during COVID" and "COVID is highlighting gender biased laws". Lastly, we discovered that users posting COVID-related misogynist tweets were causing other users in their Twitter network to become, on average, 21.9% more misogynistic. Our research supported our belief in the urgent need for effective counter-measures and we redoubled our efforts to implement targeted, innovative programmes that sought to mitigate the harmful effects of online misogyny and gender-based hate speech.

Counter-measures

I have been privileged to lead Mythos Labs' efforts to stem the rising tide of online misogyny in various countries around the world. In this section, I would like to highlight two of the most innovative approaches we have implemented in South Asia – information literacy and humour.

"My Power" trainings – Building women and girls' information literacy to combat online misogyny

Despite increasingly stringent content moderation policies of social media platforms, online misogyny is unfortunately still a problem on social media. We at Mythos Labs believe it is therefore imperative that vulnerable populations, including women and girls, build their social media literacy skills in order to increase their resiliency to misogyny and to better position themselves for success in their communities. Investing in women and girls' social media literacy works, as evidenced by a significant body of research. For example, a recent study revealed that the use of digital technologies by adolescent girls was linked to improved academic performance, health status and self-esteem as well as lower rates of childhood marriage and teenage pregnancies.[4] An examination of information literacy efforts targeting young women in Pakistan revealed that digital skills have helped women strengthen their social networks, manage their daily affairs and engage in entrepreneurship.[5] A 2017 study titled "Empowering

Indonesian Women through Building Digital Media Literacy" showed that digital media literacy training provided rural Indonesian women with better social status, bargaining position and influence in village policies.[6]

The tangible impact of information literacy trainings on women and girls can also be found in Mythos Labs and UN Women's "My Power" trainings that have been delivered to over 300 women and girls across Bangladesh, Indonesia, the Philippines and Timor-Leste every year since 2017.

Our "My Power" trainings were originally developed in 2017 as an initiative to counter violent extremism. Research by the Royal United Services Institute's Emily Winterbotham and Elizabeth Pearson, conducted that same year, described how ISIS was using a false sense of empowerment to lure women recruits.[7] By preying upon the fact that women in repressive communities/households may feel powerless, ISIS claimed to offer women an empowering experience, one that would make them physically and spiritually stronger. For example, Sally-Anne Jones became a symbol of ISIS propaganda directed at women until she was killed in a strike in 2017.[8] The image of a Burkha-clad Sally holding a gun was used by ISIS in recruitment materials directed at women around the world, to project a misleading image of female strength and empowerment associated with the terror group. Of course, the unfortunate women who succumbed to ISIS's propaganda would, upon joining ISIS, be greeted by a grim reality much less idealistic than the one they had been sold. As per Winterbotham's comment in a news article, women who were sold this misleading promise "were saying, 'this is empowering for me' but the irony is, it's not going to happen once they get there".[9]

So, in 2017, our team and I designed a programme that sought to increase the resilience of women and girls to the gendered propaganda that extremist groups were disseminating. The original goal of the "My Power" trainings was twofold: to improve women's social media literacy so that they are able to identify and report ISIS propaganda, as well as to reinforce the fact that joining ISIS is not empowering. Upon receiving funding for the first round of the "My Power" trainings to be held in Bangladesh in 2017, our team selected an initial batch of participants. To do so, we partnered with local NGOs in Dinajpur and Jessore, two small-sized cities that were identified by the NGOs as having populations with lower-than-average levels of digital literacy, especially among women. We set two criteria for the selection of participants:

- Demographics: between 16 and 25 years of age. We wanted to target women and girls who were relatively new to social media, so as to ensure that we could inculcate healthy social media habits from a relatively early age. Based on a survey of 50 households in each village, we found that the average age at which young women began using social media was 16. We capped the age at 25 because we were informed by our local NGO partners that a significant share of women in these communities above the age of 25 were already married and would therefore have less time and inclination to attend the trainings.

- Interest in storytelling. We further screened for participants who were interested in storytelling, i.e. writing, dramatic arts, filmmaking, etc. This was important because a key component of the trainings would involve the participants creating their own "counter-narrative" videos, i.e. writing, acting and post-producing their own videos. In order for the participants to get the most out of these trainings and to create the most effective counter-narrative videos, it was therefore important for them to have an interest in the art of storytelling, so that they would be comfortable making and sharing videos that promoted positive narratives.

Through our local NGO partners, we administered a call-for-applications to women in both communities and identified 25 candidates in each community for two sets of trainings. We capped the training size at 25 so as to ensure that each participant could receive adequate attention and personalized assistance from the trainer and co-trainer.

This model of selecting participants using the aforementioned criteria in partnership with local NGOs was followed in all the countries where we administered "My Power" trainings over the next several years.

Over the next few years, as online misinformation and hate speech became bigger threats, the "My Power" trainings expanded to include social media literacy tips for countering these issues as well. This did not change the attendee selection process because the same criteria (being new to social media and having an interest in storytelling) still applied, even though the scope of issues featured in the trainings had expanded. Between 2018 and 2021, the expanded trainings were conducted in Bangladesh, Philippines, Indonesia, Timor-Leste and Sri Lanka, where UN Women had funding available for.

On the first day of the "My Power" trainings, participants were taught fundamental social media literacy skills with an emphasis on the following questions: how to identify extremist propaganda, misinformation and hate speech, how to protect themselves from malicious online users, how to safeguard their privacy as well as how to report online abuse on various platforms. On the second day, we trained the participants on how to write, film and edit entertaining short videos that showcased female protagonists overcoming relatable problems ranging from cyber harassment to misinformation. All videos created by participants were filmed and edited on their smartphones. The videos were then shared by the participants on their own social media handles in "unlisted" mode on YouTube, i.e. available only to those with whom the link was shared and not otherwise searchable. This ensured that the videos could be widely circulated within the target communities while giving the participants a great degree of control over who saw the videos. It also served to limit unwanted attention and/or comments that may have made the participants feel uncomfortable or targeted.

For many of our participants, uploading a video and sharing it with a limited number of friends and family members were not only thrilling, but also empowering experiences. Most of the participants had never owned their own smartphones, and had only occasionally been allowed to borrow their male relatives' phones. So, the act of using their own smartphone to create and share their own video made them feel, as one participant in Dinajpur smilingly told me, "like a superwoman".

The idea of having participants create a gender equality-themed video was rooted in the principle of experiential learning, i.e. learning by doing. Multiple studies and experiments have validated the hypothesis that experiential learning provides a range of benefits to students as opposed to static, theoretical learning of concepts devoid of accompanying practical, experience-based exercises. In his paper titled "Experiential Learning: Experience as the Source of Learning and Development", Professor David Kolb finds that learning is best conceived of as an experiential process, not as a set of outcomes.[10] Professor Scott A. Lee in a 2008 paper also found that experiential learning caused "deeper levels of learning and applications of classroom learning" with a multitude of benefits for students including increases in their ability to adapt to change and take initiative.[11]

We had the participants in our "My Power" trainings create gender equality-themed videos because we hoped this would help them internalize the lessons they had learned. By creating videos that showcased empowered women using social media literacy to overcome relatable threats, we aimed to reinforce the key learnings of the trainings in a way that would ensure the participants retained them well into the future.

Our hopes were largely fulfilled. Post-training surveys revealed that participants reported increased confidence in their abilities across all topics as a result of the trainings, indicating significant gains in knowledge. The mean levels of confidence among participants before and after the trainings improved significantly in every key category. As a result of the training, participants grew 33% more confident in their ability to identify the difference between hate speech, misinformation and disinformation. They also felt 43% more confident in their ability to report hate speech and misinformation on social media. Participants grew 25% more confident in their ability to write an original script for a video and 39% more confident in their ability to edit a video. While the increase in confidence levels was promising, we were aware of the risk of optimism bias influencing participants' responses to the survey questions.

We also measured long-term changes in behaviour as a result of the trainings. Six months after each training, a follow-up survey was administered to all participants to test sustainability and long-term impact. The results were encouraging – 62% of all participants reported having made at least one more gender equality-themed video on their own in the six months after the training. A 21% reported having made at least two videos and 7% had made over three videos in the same time period. This demonstrated that the trainings not only taught relevant skills and techniques to the participants, but also instilled in them a desire to continue making positive content that countered gender stereotypes.

The sustained impact of social media literacy trainings on the participants' ability to identify and report hate speech was also visible in a difficult yet empowering incident involving one of the training participants from Bangladesh. A few weeks after attending a "My Power" training, a participant named Raunak found herself the target of hate speech online. During the training, she had made a video about a woman pursuing her desire for higher education against all odds. After uploading it onto Facebook, she initially received only positive feedback from friends and well-wishers on Facebook. Within a few days, however, a Facebook user unbeknownst to Raunak shared her video accompanied by a misogynistic and derisive comment.

Raunak immediately applied her learnings from the training and followed the correct procedure to report the user's post on the Facebook platform. The post was taken down by Facebook within a few days. Raunak told us she had encountered similar hate speech in the past on public platforms but, since she never knew how to report it, her response had thus far to disable comments or limit her online activity altogether. This time, however, armed with the tactical skills she gained in her training, Raunak was able to protect her mental health without having to limit or halt her online activity.

This incident underscored the fact that even a basic understanding of social media literacy can significantly impact how safe women feel online. Reporting hate speech on Facebook may not seem complicated, but many users, especially those in rural communities with limited exposure to digital technology, are not aware of how to perform such procedures online. This incident also highlighted the fact that women who do not have high levels of social media literacy can feel like they have no alternative but to limit their digital activity, so as to avoid unwanted contact from misogynist or malicious users. This can have devastating consequences for gender equality, since spending time online is increasingly important for educational and professional success.

Countering viral lies with viral truths – Using humour to counter misogynist narratives

Research suggests that humour is an effective means of countering harmful narratives and influencing opinions. In "A Funny Matter: Toward a Framework for Understanding the Function of Comedy in Social Change", Professor Caty Chattoo describes how comedy can be levered to bring about social change.[12] In "All Joking Aside: A Serious Investigation into the Persuasive Effect of Funny Social Issue Messages", Robin Nabi, Emily Moyer-Guse and Sarahar Byrne demonstrate that respondents who find a message funny are also more likely to pay closer attention to the message, and they are less motivated to argue against its substance.[13]

We at Mythos Labs were already familiar with the effectiveness of using comedy to bring about social change, as evidenced by the success of "I Want to Quit ISIS",[14] a comedic video we had produced in 2017 that countered the narrative of ISIS. This video went on to amass over 2 million cumulative views on YouTube and Facebook, glowing reviews in the press and thousands of positive comments from viewers, some of whom credited the video with helping them develop a more nuanced and tolerant worldview. So, beginning in 2018, we decided to apply our approach of using comedy towards combatting gender stereotypes and online misogyny as well. Since 2018, Mythos Labs has partnered with some of Asia's most popular female comedians, musicians and influencers to create "edutaining" videos that use humour to counter misogynistic narratives. The videos touch on a range of issues including combatting gendered narratives used by extremists in Southeast Asia, highlighting the pay gap in South Asia, and satirizing the stereotypical representation of women as damsels in distress in popular Malaysian movies. We have created 13 gender equality-themed viral videos, all hosted on the social media accounts of the entertainers themselves so as to guarantee a wide reach among local young people.

Our videos have attracted significant attention and engagement. The 13 videos we shared amassed over 6 million views. To conduct a sentiment analysis of viewer comments, we used our Natural Language Processing (NLP) algorithm to classify each comment the videos received as either "positive", "negative" or "neutral". A "neutral" comment was one that did not deliver a judgement on the video, for example, a comment saying "Who is the actor in the first scene?" is scored by the algorithm as "neutral". Factors that the algorithm used to classify comments include the meanings of words featured in the comment, as well as types of emojis used in the comment (e.g., happy face or angry face, etc.). We found that 96.2% of all video comments were positive, 3.4% were neutral and only 0.4% were negative. Of course, comments are not the only indicator of how people feel when watching a video, but the overwhelmingly positive numbers suggest that audience reactions to the videos were likely favourable.

Where appropriate, we also conducted focus groups to better understand unexpected audience reactions. For example, one video we made was titled "Brainwash",[15] starring Bollywood actress Aahana Kumra as well as the popular South Asian comedy group "East India Comedy". A satirical commercial for a women's beauty product called "Brainwash", the video highlighted how gender stereotypes are reinforced by various elements in society, including terrorist groups.

The response to Brainwash was overwhelming and mostly positive – within a few days, it received over 400,000 views on Facebook and YouTube combined, as well as numerous flattering mentions in press articles in India.[16] Over 90% of the comments on the videos were positive, characterized by supportive messages from women and men alike. Many agreed that the video highlighted important topics and prompted a much-needed discussion on gender stereotypes in Indian society. However, a small but vocal minority of viewers did express their displeasure at the video, in particular at the video's mention of the gender pay gap. This issue proved to be quite a hot button, prompting angry comments from several male users, one of whom used the term "feminazis" to describe the makers of the video. Rather than responding to the comments, the creators decided to ignore them. This, they explained to us, was based on their years of experience as social influencers dealing with trolls. "If you respond to the comments, it's like opening a can of worms. No matter how civil your tone or noble your intentions, angry trolls will see this as an opening to respond even more aggressively and rudely", one of the writers told me.

We decided to conduct a focus group to try and understand the reactions of male viewers to this video. We interviewed 20 local men and 20 local women who had seen the video. All 20 women said they enjoyed the video and that it effectively raised important issues in an entertaining manner. Eighteen of the men agreed but two of them disagreed, citing the joke about the gender pay gap as the worst part of the video. When asked why, they insisted that the pay gap was "fake news", nothing more than a false idea perpetuated by feminists seeking to make more money, even if it came at the expense of men. Even though "Brainwash" raised several issues related to gender inequality – from marital roles to body image issues to gender stereotypes used by violent extremists – the one that consistently elicited the strongest negative reactions from a small but vocal minority of male viewers was the gender pay gap. To these viewers, the economic consequences of true gender equality seemed to be much scarier than cultural or political ones. This was an important insight that we are incorporating into future videos in the region.

Conclusion

After five years of travelling across Asia to design and implement programmes that aim to counter gender-based stereotypes and misogyny, I am convinced that we have only scratched the surface of what is possible. With the advent of new technologies such as the metaverse and increased proliferation of social media, practitioners such as myself have an exciting new canvas on which to design increasingly imaginative and effective programmes that meet young generations where they are. At the same time, the constantly evolving nature of how, where and in what manner we communicate will also give rise to new challenges related to online misogyny and gender-based hate speech.

We must accept that there will always be those resistant to changing deeply held beliefs. For example, the reaction to Brainwash brought to light the immense challenges associated with countering ingrained misogynistic beliefs. It was clear that while the majority of viewers appreciated the message of Brainwash, there was a small group of stubborn viewers who were so convinced of the validity of their sexist views that no amount of creative storytelling or appeal to facts would likely dissuade them.

This is where early-childhood education can play a critical role. Instead of waiting until misogynistic beliefs are deeply rooted in the minds of men and/or women, we must incorporate gender-sensitive perspectives in the curriculum of young children. This will help prevent the establishment of deeply held sexist views in both men and women, rendering them more amenable to counter-messaging and capacity building. To do so, governments, educational sectors and civil society must partner together to create a truly holistic approach to gender sensitization of large populations. We must expand our programming to target not only young women, but also older age groups and male members of communities, who can also be vulnerable to and/or perpetrators of online misogyny and hate speech.

Technology companies must also bolster their efforts to moderate online misogyny and gender-based hate speech. While content moderation for online platforms has traditionally been more focused on removing misinformation and extremist messaging, the same level of resources and public scrutiny must be applied towards combatting online misogyny as well.

With more women coming online than ever before, I am optimistic that practitioners and governments will be inspired by the success of innovative programmes and undertake bold new ideas to counter the ever-evolving threats facing women online.

Notes

1 Mythos Labs, "Examining Linkages between COVID-19 and Online Misogyny/ Hate Speech Directed at Women in Asia", (2020), https://mythoslabs.org/wp-content/uploads/2021/09/Linkages-Between-COVID-19-and-Online-Misogyny-Mythos-Labs.pdf

2 ibid.

3 Ibid, p. 3.

4 Salima Meherali, Komal Abdul Rahim, Sandra Campbell and Zohra S. Lassi, "Does Digital Literacy Empower Adolescent Girls in Low- and Middle-Income Countries: A Systematic Review", *Frontiers in Public Health* 9, (2021), https://doi.org/10.3389/fpubh.2021.761394.

5 Sohima Anzak and Aneela Sultana, "Social and Economic Empowerment of Women in the Age of Digital Literacy: A Case Study of Pakistan, Islamabad- Rawalpindi". *Global Social Sciences Review* 5, no.1 (2020): 102–111. doi: 10.31703/gssr.2020(V-I).11.

6 Fiona Suwana, "Empowering Indonesian Women through Building Digital Media Literacy", *Kasetsart Journal of Social Sciences* 38, no. 3 (2017): 212–217, https://doi.org/10.1016/j.kjss.2016.10.004

7 "ISIS Luring Women with False Sense of 'empowerment': Report", *The Economic Times,* 6 August 2017, https://economictimes.indiatimes.com/news/international/world-news/isis-luring-women-with-false-sense-of-empowerment-report/articleshow/59941234.cms; Elizabeth Pearson and Emily Winterbotham, "Women, Gender and Daesh Radicalisation: A Milieu Approach", The RUSI Journal, 162, no. 3 (2017): 60–72.

8 Ewen MacAskill, "British Isis member Sally Jones 'killed in airstrike with 12-year-old son'", The Guardian, 12 October 2017, https://www.theguardian.com/world/2017/oct/12/british-isis-member-sally-jones-white-widow-killed-airstrike-son-islamic-state-syria

9 "ISIS Luring Women with False Sense of 'empowerment': Report", 2017.

10 David A. Kolb, Experiential Learning: Experience as the Source of Learning and Development (n.p: Prentice-Hall, 1984), p. 272.

11 Scott A. Lee, "Increasing Student Learning: A Comparison of Students' Perceptions of Learning in the Classroom Environment and their Industry-Based Experiential Learning Assignments", *Journal of Teaching in Travel & Tourism* 7, no. 4 (2008): 37–54, DOI: https://doi.org/10.1080/15313220802033310

12 Caty Borum Chattoo, "A Funny Matter: Toward a Framework for Understanding the Function of Comedy in Social Change". *Humor* 32, no. 3 (2019): 499–523, https://doi.org/10.1515/humor-2018-0004.
13 Robin L. Nabi, Emily Moyer-Gusé and Sahara Byrne, "All Joking Aside: A Serious Investigation into the Persuasive Effect of Funny Social Issue Messages", *Communication Monographs* 74, no. 1 (2007): 29–54, DOI: 10.1080/03637750701196896.
14 East India Comedy, "I Want To Quit ISIS", 25 January 2017, video, 5:25, https://www.youtube.com/watch?v=yN9dqu6RRMg
15 East India Comedy, "Brainwash | Recommended by 9/10 Men", 27 February 2018, video, 1:43, https://www.youtube.com/watch?v=JlaQdaclIjA.
16 Shreya Mukherjee, "Honoured to Be a Part of UN Women Conference", *Hindustan Times*, 9 March 2018, https://www.pressreader.com/india/hindustan-times-gurugram-city/20180309/281599536011980

INDEX

Italicized and **bold** pages refer to figures and tables respectively, and page numbers followed by "n" refer to notes.